Saving Butterflies

This book is dedicated to the late Patrick Allfrey without whose generosity it could not have been published.

Butterfly Conservation wishes to express its thanks to all those who have assisted in compiling this Guide. Very special thanks must go to the following for their extensive comments and suggestions about the text of the book; Matthew Oates, Caroline Steel and Alan Stubbs. Many others have helped in the writing and production including Jim Asher, Nick Bowles, Martyn Davies, Kathy Gordon, Ian Hardy, Greg Herbert, Gail Jeffcoate, Ian Loveday, Peter Newton-Lewis, Andrew Phillips, Gill Pilgrim, Ken Willmott, George Yorke and Kathy Yorke.

Illustrations by Valerie Baines, Brian Ceney, Les Hall,
Ian Loveday and Rob Still.

Photographs by David Dunbar, Simon Glover, Bill Gerrard,
Greg Herbert, Anthony Hoare, Ian Loveday and Ken Willmott.

Distribution maps are redrawn from those in *Butterflies of Great Britain and Ireland* Vol. 7 (1990) with the kind permission of Harley Books.

Typeset by Quetzal Communications
Designed and produced by Meridian Design Associates
Printed in the UK by ABS Litho Ltd.

Published by The British Butterfly Conservation Society Ltd.
PO Box 222, Dedham, Colchester, Essex CO7 6EY
Telephone (and Fax) : 0206 322342
Registered Charity Number 259347

ISBN 0 9512452 9 5 Paperback
ISBN 0 9522602 0 4 Hardback

THE VALUE OF OUR BUTTERFLIES

Butterflies in Britain are a part of our natural heritage and the most popular group of insects. Butterflies are completely harmless as they do not bite or sting. In Britain only the larval stages of the Large and Small Whites may cause minor damage to some vegetable crops.

They are easily recognisable because of their intricately patterned and colourful wings. Their fragile beauty symbolises the warmth and peace of summertime and the mysteries of their transient life-cycle are an irresistible fascination. Some cultures attribute deep spiritual or religious significance to butterflies.

For these aesthetic and sentimental reasons alone, butterflies deserve protection. However, their scientific and practical value is of far more significant importance. Here are the reasons why:

THE HEALTH OF THE COUNTRYSIDE

Butterflies are prime indicators of the health and condition of the countryside. If butterflies are living and breeding the land will almost certainly be environmentally sound. However, they are highly sensitive to changes which are unsuitable or hostile to their life-cycle requirements.

Being day-flying creatures they are easy to see. Regular monitoring of butterfly populations is proving a reliable method of identifying adverse changes in the countryside; and will be helpful in following climatic variations. The loss of butterfly colonies or reduction in numbers is usually a reflection of urban, industrial or agricultural pressures.

POLLINATORS

Butterflies are important for the pollination of flowers. Their long tongues enable them to take nectar from certain plants upon which most other insects cannot feed.

BALANCE OF NATURE

Butterflies, and especially caterpillars, are a major food source for birds and other animals. They also act as host to regulating parasites, thereby supporting many species and contributing to the balance of nature.

EDUCATION

Educationally, butterflies are of interest at every level of study. From an early age children find them a fascinating and absorbing subject. The amateur naturalist will derive endless pleasure from observing butterflies and, nowadays, an involvement in conservation activities. At a more advanced educational level they have many attributes for understanding ecology and environmental principles.

RESEARCH

Scientists have demonstrated the significance of butterflies in research and ecology. Many evolutionary and natural processes have been discovered including major advances in chemistry, biology and genetics.

ART

Butterflies are a source of inspiration to poets, writers, artists and photographers. Happily the advance of photographic technology over the last twenty years has meant the camera largely replacing the collector's net.

FOCUS OF AWARENESS

The intrinsic attractiveness of butterflies is an excellent focus for increasing peoples' awareness of conservation issues.

A hostile environment where butterflies will be few and far between. Intensive agriculture, roadbuilding, land development, pollution and rubbish seriously threaten or destroy habitats.

THE BUTTERFLY LIFE-CYCLE

The life cycle of a butterfly from ova (egg) to larva (caterpillar) to pupa (chrysalis) and to imago (adult butterfly) is highly complex and differs greatly between species. Conditions must be within the tolerance range at all stages of development, otherwise it will die. At certain points of growth, conditions may be critical, whereas at other times there will be less danger.

The suitability of a habitat and its environment for any butterfly species can be assessed only with a detailed knowledge of its life cycle and the conditions under which it will develop. It should also be remembered that butterflies, unlike some plant seeds, cannot remain dormant so the life cycle must be repeated from one year to the next.

The life-cycle and key factors for success at each stage are:

OVA

- Acceptable location, siting density and stage of maturity of foodplant for egg laying
- Butterflies may lay eggs singly or in clusters on the foodplants, others may lay in the vicinity or drop their eggs in flight
- Maturing and hatching without disturbance.

LARVA

- Supply of foodplants coinciding with the caterpillar feeding periods
- Skin moulting - 4 to 6 times, depending on species
- Resting safety - protection from predator and parasite attack through camouflage or concealment
- Weather - some species overwinter as larvae, others need especially warm conditions
- Dispersal - some species live in nests or clusters but may break away later to feed in smaller groups or singly.

PUPA

- Suitable undisturbed site for caterpillar to pupate - this may be on the ground, fixed to a plant stem, leaf or firm object
- Safe emergence - a butterfly usually dries its wings by hanging downward from pupa casing or nearby stem or leaf.

IMAGO

Behavioural patterns of mature butterfly may include:

- Dispersal or migration
- Searching for micro-habitats with higher temperature than surrounding areas
- Setting up of 'territories' patrolled singly or colonies
- Feeding on flower nectar, moisture, salts and other nutrient sources
- Resting, sun-basking, night-time and bad weather roosting and sheltering
- Courtship and mating
- Hibernation in winter to re-emerge the following spring.

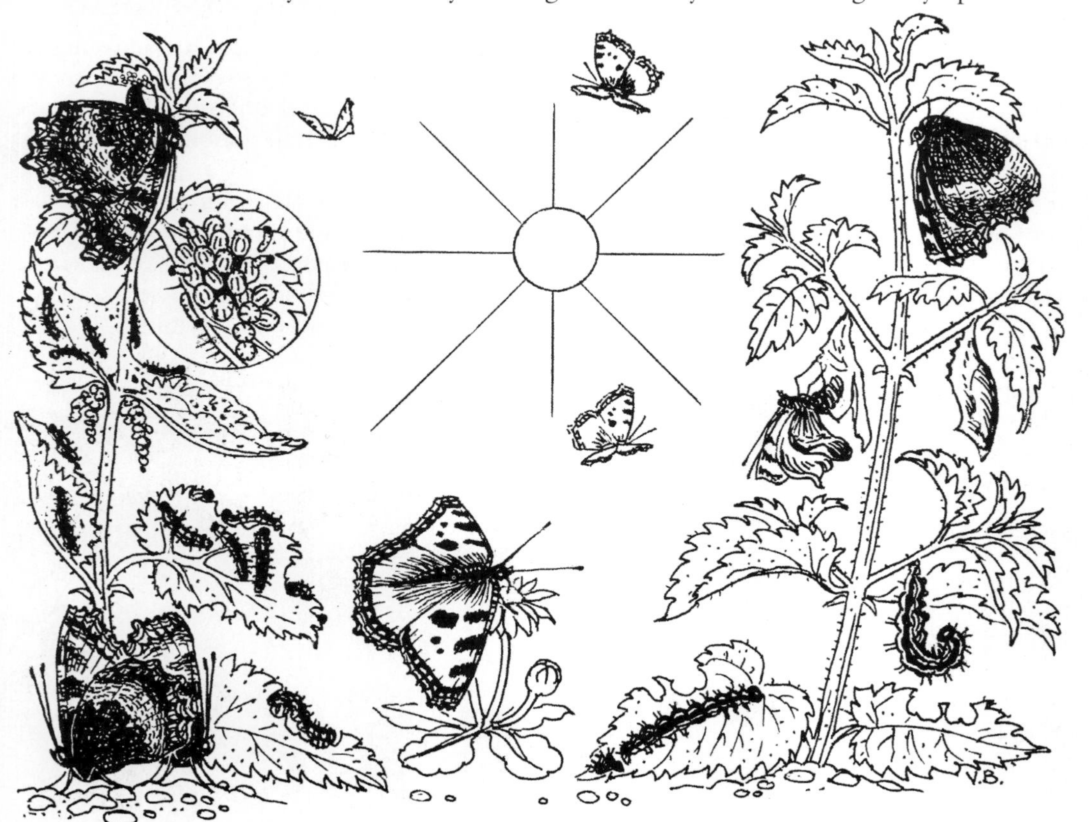

BUTTERFLY ECOLOGY

Butterflies are sensitive to many environmental factors, and the reasons for their success in one locality and absence from another are often complex.

RANGE & DISTRIBUTION

Climate and geology combine to produce an endless pattern of weather and soil conditions which may or may not be suitable for butterflies. Being cold bodied creatures they each need particular degrees of heat and light intensity to function. The hot and humid tropics are the most favourable for butterflies, with a gradual reduction in species in cooler parts of the world.

Butterflies in Europe (about 400 species), and especially in Britain (55 species), tend to be near the edge of their range. This is illustrated by the fact that many more butterflies are found in the mild southern parts of the British Isles than in the north where it is colder and tends to have less sunshine.

Some species have very exacting requirements for their life cycle and are, therefore, only found in specific locations. Others are much more tolerant of differing conditions. Relatively common butterflies, such as the widespread Small Tortoiseshell or Small White, may turn up almost anywhere. Others, such as the Glanville Fritillary (Isle of Wight), and Lulworth Skipper (Dorset coast), just survive at the northern edge of their range. A few are adapted to a much harsher climate - for example, the Mountain Ringlet (mountainous areas of the Lake District and Scotland) only flies at an altitude above 350 metres. It follows that many butterflies common in the southerly regions are much rarer further north to a point at which they cease to be found at all.

CLIMATE

Sunshine, temperature, humidity and rainfall are all key climatic elements influencing butterfly biology and ecology. Their effect on habitats is crucial to the development and behaviour of butterflies at all stages. The caterpillars of some Fritillary species need to sun bask as part of their growth process.

The duration and intensity of light and heat will affect the speed of growth and general activity, increasing in hotter weather. Below average temperatures will slow activity and may cause a reduction in numbers by shortening the female egg development and laying periods. Late frosts or freak hail or snow storms in summer can be disastrous and

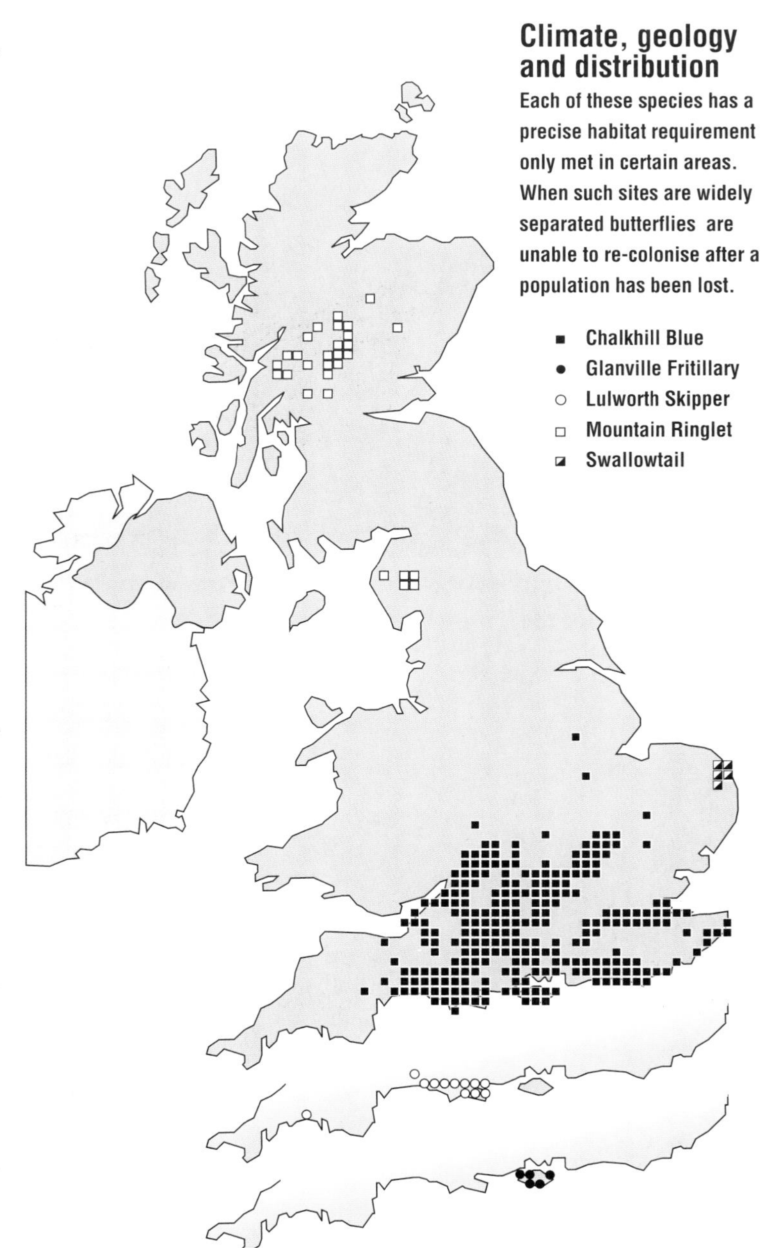

Climate, geology and distribution
Each of these species has a precise habitat requirement only met in certain areas. When such sites are widely separated butterflies are unable to re-colonise after a population has been lost.

have been known to wipe out whole colonies of butterflies.

However, during winter, severe low temperatures can be tolerated by many species and may even be beneficial in controlling disease, bacteria, parasites and predators.

Extreme climatic conditions can have a dramatic effect on the abundance and distribution of butterflies. The exceptionally hot summer of 1976 resulted in an explosion of numbers. Unfortunately, towards the end of the season, a lack of rainfall decimated some foodplants, starving the larvae. There was a similar occurrence in 1990.

Longer – term climatic changes may mean that some butterflies will no longer find conditions favourable but others may be able to extend their ranges. Global warming, (the greenhouse effect), if it continues, may result in radical changes in butterfly distribution. with an increase in numbers and species in Britain. On the other hand plants may be adversely affected if the climate becomes too dry.

New roads, like this one at Twyford Hill Down in Hampshire, take an enormous amount of land and may destroy irreplaceable habitats and butterfly colonies.

Loss of open heathland to industrial development and housing estates is a particular concern in the South-East, as in this picture from the New Forest.

GEOLOGY

Geology underlies the landscape and interacts with the climate. Rock and soil types combine with altitude, contour and aspect to dictate which plant and animal life will best flourish. Butterflies have adapted to nearly all types of landscape - some favour low-lying flat country whereas others live only in hilly or high mountainous areas.

Aspect and slope influence the amount and intensity of sunlight reaching the land surface - they may also give shade and wind shelter. Soil temperature may be critical for the right breeding conditions of some species - for example, the Adonis Blue is exclusively found on warm, south facing chalk slopes.

Plant life depends on soil type, moisture, aspect, climate and competing species. For feeding and breeding, butterflies have to search out the preferred foodplants growing and flowering to coincide with the appropriate stage of life-cycle development.

The distribution of the larval foodplant often determines the areas where a butterfly is found - especially those with migratory tendencies. The distribution of the Brimstone in Britain follows the same pattern as its foodplants, Purging Buckthorn and Alder Buckthorn.

MAN'S INFLUENCE

Man's activity on the land is the final and all too often decisive factor - there is virtually nowhere that can truly be said to be natural or untouched countryside. At some stage the land will have been disturbed by clearance, cultivation or building. We do, therefore, owe much of our butterfly population to the type of countryside that has evolved over the last few thousand years as a result of man's influence.

Equally, it cannot be expected that the country will remain static. There are constant changes either through the processes of nature or interference by man's activities. Indeed, our concern for the survival of butterflies arises from an awareness of the scale and rapidity of these changes, especially in land use and management practices.

HABITAT TYPES

GRASSLANDS FOR BUTTERFLIES

Almost all grassland in Britain has been created and maintained by man's activities. After the last Ice Age woodland vegetation became established everywhere except on land higher than 3000 feet, and a few small areas inhospitable to trees such as steep, unstable slopes. When the trees were cleared by man and prevented from returning by grazing animals, grassland became established.

Grassland is usually described as acid, neutral or calcareous, according to the nature of the underlying rocks. Acid grassland occurs mainly in the uplands and west of Britain and contains distinctive grasses and a rather limited range of other plants, as these conditions are inhospitable.

Calcareous grassland occurs on the chalk and limestone hills, mainly in the south and east. It has a far richer variety of plants and is best for butterflies. The distribution of some species is restricted to this geology because their foodplants only grow in chalky soil, while others survive because the south facing slopes with short turf provide a warm micro-climate. Often old chalk quarries and ancient earthworks create excellent butterfly habitats.

Variations of soil type, drainage, altitude and climate make grasslands an extremely varied habitat with many different types of flora and fauna.

The land in the background has been heavily grazed by sheep and is dried out after a hot summer. Flowers and plantlife still flourish in the protected fenced field in the foreground.

Threats to traditional grassland

In most counties up to 90% of unimproved grassland has now been lost. The main threats to ancient meadows, pasture and natural grassland are the destruction or diminution of their plant life by the following:

- **Agricultural improvement by ploughing or rotovating to plant crops or re-seeding with modern high-productivity varieties of grass: in either case, the resultant plant life will not support any butterflies. Rye grass leys are useless for butterflies and associated wildlife.**
- **Overgrazing: heavy grazing by farm stock or in some instances by wild animals, especially rabbits, will reduce the diversity of plant life.**
- **Drainage: the lowering of the water table by ditching, underdrainage or ground water abstraction leading to the drying out of the topsoil. This destroys the special quality of the site reducing the variety of plant life and humidity.**
- **Insecticide spray and drift: Lowland farm crop spraying with insecticide often threatens adjoining butterfly grassland and hedgerows.**
- **Herbicides and fertilizers: these are designed to control or modify plant growth but invariably reduce the number of plants useful to butterflies. They may also allow a few of the most vigorous to become dominant at the expense of others.**
- **Lack of management - neglect: when grassland is left unmanaged it will rapidly revert to coarse grass, scrub and trees. The latter will dominate, leaving few, if any, of the original plants or grasses.**
- **Total destruction: building & road development obliterates grassland.**

Continuous use for many centuries as pasture for grazing and meadows for hay allowed many species of plants and insects to thrive in grassland. Over four hundred different flowering plants may be seen on these unimproved grasslands across the country, with up to a hundred in the richest meadows. In recent decades, however, this continuity has been lost and grassland, either converted to other uses or replaced with monocultures of rye grass, which keeps the landscape looking green but does not support butterflies at all.

Nowadays many small isolated fragments of grassland remain which are of use to butterflies or other insects. These include field margins, woodland clearings and ride edges, churchyards which have not been tidied, some railway and road embankments, and remote coastal strips, cliff tops and other odd corners. The Meadow Brown will colonise relatively insignificant patches of uncut grassland even in towns and

Typical grassland butterflies

- **All grassland: Meadow Brown, Ringlet, Skippers - Large, Small and Essex, Common Blue, Small Copper**
- **Acid lowland and uplands: Small Heath,Small Copper**
- **Chalk and Limestone: Adonis and Chalkhill Blues, Silver-spotted Skipper, Brown Argus, Grizzled Skipper**
- **Damp meadows: Marsh Fritillary**

cities, so butterflies can be helped and encouraged almost anywhere.

CONSERVING GRASSLAND FOR BUTTERFLIES

The purpose of maintaining grassland is to ensure that existing nectar and larval food-plants can continue to flourish in the correct conditions, particularly for targetted species. Protecting an area is not enough and active management is essential. Also, the site may be assessed for the introduction of new plants or to re-establish ones which have died out in order to encourage other butterflies (relevant codes and practices for both plants and insects should be followed).

Each species of grassland butterfly has a preferred turf height and density of grass - this applies especially to some of the 'Blues'. So before starting any management on a site one must identify and monitor the species breeding there. This should last at least one season (see section on Transect Walk). Therefore, to sustain a variety of species there should be a 'patchwork' or 'mosaic' of differing turf heights on the site unless it is being specifically managed for one type of butterfly. Careful allowance should be made for other wildlife on the site and differing qualities and potentials sustained.

Conserving grassland can be best achieved either by grazing or cutting in small compartments. These should be planned in relation to the prevailing conditions, which will dictate the speed of growth and therefore the type and numbers of livestock to be grazed or frequency of cutting. Never risk grazing or cutting the whole of a site at once.

Burning is a third alternative, but generally this is not recommended since the insect fauna is likely to decline. The reclamation of a site will necessitate a more intensive programme of cutting or grazing. There will then be a point at which this is modified when the land is restored and merely needs maintaining.

GRAZING

Many grassland habitats owe their characteristics to the way in which animals feed. This may vary, from livestock which are tightly controlled by fencing to wandering flocks or herds over open land. Grazing in one form or another is regarded as perhaps the best way of managing grassland for butterflies and other wildlife.

Grazing removes herbage from the land but at the same time helps replace the nutrients by urine and dung. Animals grazing and trampling keep plants low and prevent build-up of dead plant 'litter' which might smother new growth and allow weaker but vital plants to

Once flower rich meadows and downland existed all over the countryside - now they are a rarity. They have been 'improved' and ruined with fertilisers, insecticides and drainage, often with government subsidies.

Sheep paths on lightly grazed downland in Dorset.

compete. Grass and plants are removed slowly by animals moving over the land in flocks and herds.

Wild animals, and in particular rabbits, if present in sufficient numbers and undisturbed, will graze swards thereby restricting plant development.

Over-grazing is a serious danger and should be avoided at all costs. Plants may be destroyed or reduced to insufficient foliage - then common weed plants may take over bare soil. It is always better to under-stock and prevent damage as more animals can be used at a later date.

LIVESTOCK MANAGEMENT

Grazing can be achieved by continuous low density rotational, seasonal or occasional grazing. The choice will depend on many factors but it is likely that a combination of these options, carefully monitored, will produce the most successful results. Different methods are likely to be suitable for certain areas especially if a large or varied site is being managed for butterflies.

Continuous grazing should be with small numbers of animals - probably 1-3 sheep or 1 cow for every hectare. A doubling of these numbers would be considered high density for management purposes but in any event the effect of grazing must be carefully watched. All animals graze selectively so it may be necessary to control them in such a way that they will eat the less favoured diet in order to remove the unwanted vegetation.

Rotational stocking means moving animals from one compartment to another under a planned programme. This method of grazing is probably the most effective provided fencing and labour resources are available and the effects of grazing are constantly monitored. It is safer to graze only on part of a site each year with a cycle of two to five years depending on conditions. Care should be taken not to have all the particular habitat requirements of one butterfly confined to any one section in case there is inadvertent damage by grazing.

Seasonal grazing is safest during winter between late September and February. It should be avoided during summer months unless it is part of a planned low density or rotational stocking. The same principles also apply to occasional grazing. Lay-off land should be available so that animals can be removed from a conservation area at short notice. Whatever system is used under-grazing is always safer to avoid irreparable damage to butterfly colonies and the land.

The use of livestock for habitat maintenance should only be undertaken in close co-operation with the landowner. The purpose and terms of the grazing should be fully understood and agreed with the stock owner. A grazing regime will involve careful organisation and a degree of flexibility, including the following:

- Area to be grazed and timing
- Types and number of livestock
- Means of control, i.e. permanent or temporary fencing or electric fencing
- Means of transport and access
- Water supply and other feeding requirements
- Regular monitoring of the grazing progress
- Movement of the fencing if grazing in sections
- Shelter from wind and rain
- Lay-off land
- Legal agreement or licences with stock owner

CUTTING AND MOWING

Cutting is not the best method of managing butterfly grassland and ideally should be used only where grazing is not a viable option.

Cutting pattern

The shape and size of the site must be identified. A pattern of cutting areas should be devised

Animals for grazing

■ **SHEEP:** most butterfly habitats can be maintained by sheep with the right stock numbers and careful timing. They are especially suitable for limestone grasslands where sheep walks (mini-terraces with bits of bare ground) create the ideal conditions for certain species. Selection of the breed and age of sheep should be appropriate to the region and land conditions. Sheep under 3 years old graze less well as their teeth have not matured enough.

■ **CATTLE:** cattle are best suited for eating longer grasses and vegetation or on sites which have not been grazed recently. The heavier weight of cattle will break up the ground and turf through trampling thereby enabling low growing plants to regenerate. However, this should be done with great care to avoid damage especially on damp ground or in wet weather. Different breeds and ages behave differently - young bullocks rush around madly, whereas mature milkers are docile.

■ **GOATS:** tethered goats are well suited for clearing small areas of tough or dense vegetation. If regularly moved they are ideal for the creation of a habitat mosaic across a site. They should not be left on one site too long as they can create a 'bowling green' effect. Goats are excellent browsers and best for dealing with scrub.

■ **PONIES AND HORSES:** these are highly selective grazers but can produce excellent grassland mosaics. They should be used only in low density numbers to avoid over-grazing. One or two animals per hectare will usually be sufficient. Beware of nervous or highly strung animals which are hard to control.

■ **RABBITS:** wild rabbits feed selectively and may contribute some benefit towards maintaining a butterfly site. If they become too plentiful grassland will be close cropped and some plants eradicated altogether. With myxomatosis a lessening control on numbers, it may be necessary to keep rabbits in check by artificial methods. They can also be a menace by disrupting planned habitat management.

to form a loose mosaic so that only some sections are cut at any one time or during any one year.

The illustration below shows a typical pattern, the main feature of which is fragmentation thus avoiding too great an area being affected. If butterflies are flying, they must be able to move to an uncut section in close proximity. But it is better not to cut during the flight and laying seasons altogether, except where a small mosaic of summer-cutting is carried out on part of a site.

Timing

Frequency of section cutting will depend on growing conditions and must be planned in advance. A rotation of one or two cuts per annum will suit most grassland. Areas of poor or thin topsoil especially on steep slopes supporting only slow growing vegetation should be left much longer - some grassland may be left for years on end. Frequent monitoring will determine the effects of cutting and variations in the timing are inevitable. Weather conditions are unpredictable so there will be a continuous process of trial, error and adjustment.

Generally it is considered best to cut grass during early spring (not after April) or mid-autumn (not before mid-September), and preferably later when caterpillars and butterflies no longer need nectar flowers. Autumn cutting on sites with heavy soils will gradually reduce the plant variety so a mosaic pattern should be cut in the summer.

Hay crops may have to be taken but the longer they are left the better it will be for the butterflies. Normally summer haymaking does not suit butterflies so should only be a priority where this is a long established farming practice.

Mowing along tracks and paths should be on a rotational basis to form a strip of mosaic on either side. Leave small uncut patches where dead flower and seed heads offer roosting places for butterflies and over-wintering shelter for other insects.

A butterfly field: south-facing aspect or slope, sections cut or grazed in rotation, varying turf lengths and density, a profusion of grasses, wildflowers and herbs, clumps of bramble and shrubs, nettlebeds, a network of paths with bare earth and stones, anthills, and tree shelter-belt.

Cutting methods

A coarse cut of about 8-10 cm is best for butterflies but a shorter cut may benefit some plants or a few butterfly species so the cut height can be varied by choice. Gang mowers should be set high and hay mowers similarly, although this will depend on cutting conditions. Short, lawn length mowing should be avoided at all costs. The least damaging method of cutting is by hand-scything. Usually this is not practical on any great scale so mechanical mowing will be necessary.

In small or confined areas hand-scything is the safest method of cutting. The use of an electric or petrol-driven strimmer may be convenient but take care not to cut too close.

Cylinder or hover-type lawn mowers are not recommended, as these give too short a cut and are likely to kill insects in the early stages of development.

Leave small areas of taller grasses and shrubs. Brambles provide nectar flowers and shelter. Avoid the destruction of anthills especially on sites with 'Blue' butterflies. Clear all cut or dead material from sites to prevent choking of new plant growth and the fertilising effect on soil.

Where there has been invasion of scrub, saplings or trees, the use of long-handled secateurs or bow saws may be necessary. Stumps should be cut as near as possible to the ground and chemically treated by painting, not spraying, to prevent re-growth. Chain saws should only be operated by people with professional training and protective clothing.

Regeneration areas

Areas of a few square metres may be scraped bare of top-soil at regular intervals of, say, every twenty five metres along woodland rides and paths, hedgerows, in open fields or other habitats. A similar effect may be achieved by rotovating. These will create a seeding area for regeneration and greater variety of plants. Mole casts may produce a similar effect, although on a smaller scale. Bonfire sites will kill ground cover leaving ash and bare soil which will regenerate slowly. Scraping off may expose a different sub-soil which will support other valuable plants.

Cut material

As a general rule, cut material should, if possible, be removed, although this may be labour intensive. Rake and stack grass and light vegetation for burning or removal. Older scrub may be dragged into a wood, but this should only be on a small site to avoid unnecessary ground damage. If unwanted cuttings cannot be disposed of then, if carefully stacked, they may provide a habitat refuge for other wildlife whilst decomposition takes place over several years.

Controlling bracken

Bracken is difficult to control, but if allowed to become dominant will obliterate ground cover. Cutting and spraying is the only effective means of control, but this is expensive and labour intensive. The regular cutting of bracken along the edges of paths and tracks will enable smaller and weaker plants to grow and provide nectar and caterpillar foodplants.

Removal of dead bracken will expose bare earth and is another method of creating regeneration areas. A word of warning, avoid skin contact with sap and inhaling smell of freshly cut material.

Wild seed harvesting

A problem causing increased concern is the commercial harvesting of wild seeds. Conservation sites are prime targets as they are rich in wildflowers and grasses ideal for recreating grassland with a varied plant life. Large vacuum type machines hoover up the seeds but also suck up the insects which are killed. The quality of both the flora and fauna on the site is thereby diminished very quickly. If wildflower seeds are sown they should be British stock and preferably grown locally.

SET-ASIDE

The food surpluses being grown by farmers in the '80s led to government schemes to take some land out of production. Under 'Set-Aside' land had to lie fallow for five years subject to tight rules on management.

At first sight this might appear good news for grassland butterflies. Unfortunately, the organisation of Set-Aside has not been geared towards conservation and any benefits have been very limited.

It was the old traditional, low-intensity method of farming which allowed wildlife to flourish, not simply abandoning the land. Under Set-Aside restrictions there is no impover-

Above: conservation volunteers help to clear scrub on many sites.
Left: a large area of Dunstable Downs has been cleared of invading hawthorn and is now lightly grazed by sheep.

Grassland action summary

- Protect ancient unimproved plant-rich grassland
- Identify species present on site and exactly where they breed
- Avoid ploughing, rotovating or draining
- Avoid use of fertilizers, herbicides and insecticides
- Encourage diversity of flora and fauna
- Prevent development of coarse grasses and scrub
- Use traditional methods of grazing or haymaking
- Allow tall grasses for roosting and breeding
- Leave scrub and bushes in small patches for shelter
- Manage land to preserve balance best for existing land features
- Plan a conservation programme and monitor regularly
- Be cautious of proposals for wild seed harvesting

Both fields have been ploughed. Sheep now graze on a single grass crop. A fence replaces a hedge and remaining hedge tightly cut. Little chance for butterflies to breed here.

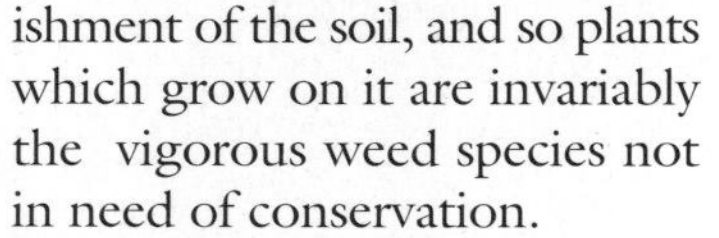

ishment of the soil, and so plants which grow on it are invariably the vigorous weed species not in need of conservation.

Set-Aside rules have been frequently amended – and now there is the prospect of a new system being implemented. With greater awareness of conservation issues it must be hoped that management regimes will be more sensitive to the needs of wildlife.

Even so, uncultivated land will be colonised by some nectar flowers such as thistles and caterpillar foodplants. This in turn will attract some butterflies to feed and breed in these abandoned fields, but permanent colonies cannot be established if the fields are returned to production after 5 years.

GRANTS FOR FARMERS AND LANDOWNERS

Financial help for conservation and diversification is available under a wide range of schemes. These are operated by the Ministry of Agriculture, Fisheries and Food, the Countryside Commission and various other organisations. Before starting any conservation work it is essential to investigate the possibilities for obtaining grants. Rules are, needless to say, complicated and constantly being changed.

In addition to Set-Aside, other important schemes which may offer opportunities for funding grassland management are the Countryside Stewardship, Environmentally Sensitive Areas and Nature Conservation Grants.

FURTHER READING

- *The Management of Chalk Grassland for Butterflies (1986, NCC)* is an easily readable report which will prove invaluable to butterfly conservationists working on all types of grassland, not just chalk. It contains practical guidance on grassland management which can be adapted to most situations.
- *The Conservation of Meadows and Pastures (1988 NCC/ English Nature)* and *The Conservation of Cornfield Flowers (1989 NCC/English Nature)* are both useful booklets
- *The Management of Grassland and Heathland in Country Parks (1977, Countryside Commission).* This book also contains much useful practical and technical information including case studies relevant to butterfly habitat conservation.
- *Conservation and Diversification Grants for Farmers (1992, MAFF)*

Set-Aside does give limited opportunity for butterflies. Grassland species like Meadow Brown will colonise quickly; others take many years before the conditions are right. Fast-growing weeds, such as thistles, offer nectar to the Small Tortoiseshell, Peacock and other mobile species. Management rules for Set-aside must be adjusted to encourage, rather than destroy, wildlife.

HEDGEROWS AND FIELD MARGINS

"And when ye reap the harvest of your land, thou shalt not reap the corners of thy field, neither shalt thou gather the gleanings of thy harvest: thou shalt leave them for the poor and the stranger."

Britain still has about 650,000 kilometres of hedgerows. With an average width of 2 metres this represents over 100,000 hectares of land. Often they sustain the only natural vegetation on otherwise 'desert' farmland. Regrettably, in many areas hedges have been removed, leaving a featureless landscape and a highly diminished wildlife population.

The hedgerow, shelter belt and woodland edge and their margins create an ideal habitat in which many species of butterflies feed and breed because of the profusion of plant, shrub and tree life. They also offer shelter from the wind, sun traps for basking, and shade for some species that prefer partial sunlight.

Contrary to common belief, hedgerows are nearly all man-made and originally designed to control and protect live-stock.

Thousands of miles of hedgerow have been destroyed.

A rough unkempt hedge and field margin provide a rich wildlife haven.

They contain bushes or trees such as hawthorn, blackthorn, buckthorn, sallow, holly, dog rose and bramble. Oak and elm in hedgerows provide useful foodplants for many species of moths. Trees and undergrowth will depend on the soil, age and locality of the hedgerow. Many forms of wildlife will flourish in hedges, especially birds, game-birds, small mammals and all sorts of insects.

Natural streams or ditches along hedges encourage the growth of plants which prefer wetter conditions, such as lady's smock, foodplant of the Orange-tip. Some butterflies, such as the Ringlet, prefer damper grassy habitats.

Hedgerows act as crucial links or corridors by facilitating the easier movement of butterflies from one habitat to another. They afford protection and sustenance when crossing open countryside devoid of shelter and nectar sources.

Sections of hedgerow are often used as 'territories' by individual male butterflies.

Threats to hedgerows

- **Removal of hedgerows for increased crop yields and easier use of agricultural machinery**
- **Close ploughing to field edges destroying margins**
- **Low flat top and close side cutting of hedges**
- **Spraying and spray drift of insecticides, herbicides, and fertilizers**
- **Stubble burning**

Butterflies may be found hibernating in hedges. The Brimstone is well camouflaged amongst ivy leaves, as is the Peacock on tree trunks or in hollows.

CONSERVATION

Sympathetic management of hedgerows, their margins and adjoining field edges is of crucial importance to butterflies. Recent research by the Game Conservancy has shown that not spraying crops along field edges automatically increases the abundance of butterflies and the variety of species. If these unsprayed strips, known as headlands, are also left fallow the benefit for wildlife is even greater. Similar management principles may also be applied to margins or verges along fences, walls and other field boundaries.

ASPECT

Most butterflies prefer sunshine. Hedgerow and headland conservation activities for butterflies should be concentrated on south-facing sides or the aspects receiving the most sun during the warmest part of the day. Conversely, north-facing shaded edges of woods or high hedgerows will be cooler and not so attractive to butterflies or for breeding areas, even though many moths and other insects thrive under these conditions.

This neglected hedgerow and ditch have been completely grazed out by cattle.

Crop spraying has damaged the base of this hedge.

Well managed field margins can support many butterflies. This one is protected from grazing cattle by an electric fence, allowing some valuable wildflower growth.

HEDGEROW MANAGEMENT

A mixed hedge is best managed by rotational trimming on a planned two or three year cycle - for some butterflies even longer. A hedge should not be cut annually. Tall hedges with a thick base will support a better variety of plants and wildlife. Close cutting and mechanical flailing should be avoided or at least used with discretion. Tapering the top in an 'A' profile is preferable to flat cutting especially for saplings. Grassy margins should not be allowed to become overgrown by hedges. Cutting should be carried out during late autumn or winter months.

Traditional methods of planting, laying and trimming hedges require considerable skill and time. Due to high costs and shortage of labour most farmers do not maintain stock proof hedges but will use wire fencing instead. Provided hedges are retained and managed in the way already described, butterflies will undoubtedly benefit.

Oak, elm, willow, sallow, birch, hawthorn and poplar will be beneficial to both butterflies and moths. Many hedgerows have lost trees and the tagging of suitable species to be left by the cutter will assist wildlife and improve the appearance. Awkward field corners may be inaccessible or left to go wild with grasses, nettle beds and bramble patches which will be alive with butterflies and managed as uncultivated grassland.

HEDGEROW MARGINS

Butterflies will greatly benefit from uncultivated grassland margins of at least two metres (wider if possible) being left along the sunniest aspect of hedgerows. Ploughing and crop seeding should allow for the widest possible 'wildlife' margins. These strips should not be allowed to become overgrown by scrub and will benefit from rotational cutting on a two to four year rotation. Where livestock are grazed hedgerow margins can be protected by temporary electric or permanent fencing a few yards in from the hedge. Such margins give butterflies and their nectar

Typical hedgerow butterflies

Brimstone, Orange Ttip, Small White, Green-veined White, Small Heath, Meadow Brown, Gatekeeper, Ringlet, Speckled Wood, Holly Blue, Small Tortoiseshell, Peacock and Comma.

flowers a chance to survive when the pasture land is heavily grazed and vegetation cropped very short.

FIELD HEADLANDS

Pesticides are indiscriminate and, apart from pest species, may eliminate or reduce butterfly and other beneficial insect populations. Herbicides destroy 'weeds' many of which provide nectar flowers or caterpillar foodplants for butterflies. (Some selective spraying may be necessary to control serious weed encroachment). **Scientific experiments show that if spraying and spray drift within six metres of field edges are avoided then insect numbers, especially butterflies, in hedgerows and margins will dramatically increase.** Whilst the effect on crop yields is minimal butterflies have been found in far greater numbers. In turn some game numbers have increased as there is a greater supply of insect food.

Spray drift is influenced by factors such as nozzle type and pressure, boom height, wind and temperature. Use a coarse spray whenever possible and set boom height close to crop. Check wind direction to avoid drift over headlands and do not spray if wind speed exceeds 4 mph.

An uncropped headland is even more beneficial as this is left without sowing and no spraying is necessary. Uncropped headlands can be any width thereby reducing surpluses. Occasional tilling every 2 or 3 years (mid-October is best), will prevent perennials taking over and allow annuals to reseed.

STUBBLE BURNING

Uncontrolled stubble burning may damage hedges and margins. It can be especially destructive to the earlier butterfly life stages and even the insects in flight may not escape. The practice has never been recommended and legislation has now banned it.

FURTHER READING

- *Pesticides: Code of Practice for Safe Use of Pesticides on Farms and Holdings* (HMSO, 1990)
- *Hedging - A Practical Conservation Handbook* (Revised edition. 1988, British Trust for Conservation Volunteers)
- *The Cereal and Gamebirds Research Project* (The Game Conservancy, Burgate Manor, Fordingbridge, SP6 1EF).

Action summary

- **Protect hedgerows from destruction**
- **Replant destroyed hedgerows with native trees and shrubs**
- **Trim hedges and cut margins in rotation. Taper hedges at top**
- **Allow extra base thickness and height growth**
- **Allow trees and shrubs to mature at selected intervals**
- **Encourage diverse vegetation**
- **Link woods and spinneys with hedges**
- **Leave inaccessible and awkward corners to grow as wild grassland**
- **Create 2 metre (or wider) margins especially on southerly aspects**
- **Avoid spraying and spray drift within 6 metres of margins**
- **Never burn stubble**
- **Reduce grain surpluses with uncropped headlands**

Field headland for butterflies: the hedge grows rough and is trimmed occasionally, the margin is a minimum of 2 - 3 metres, crop spraying is avoided within 6 metres of the field-edge and butterflies will flourish.

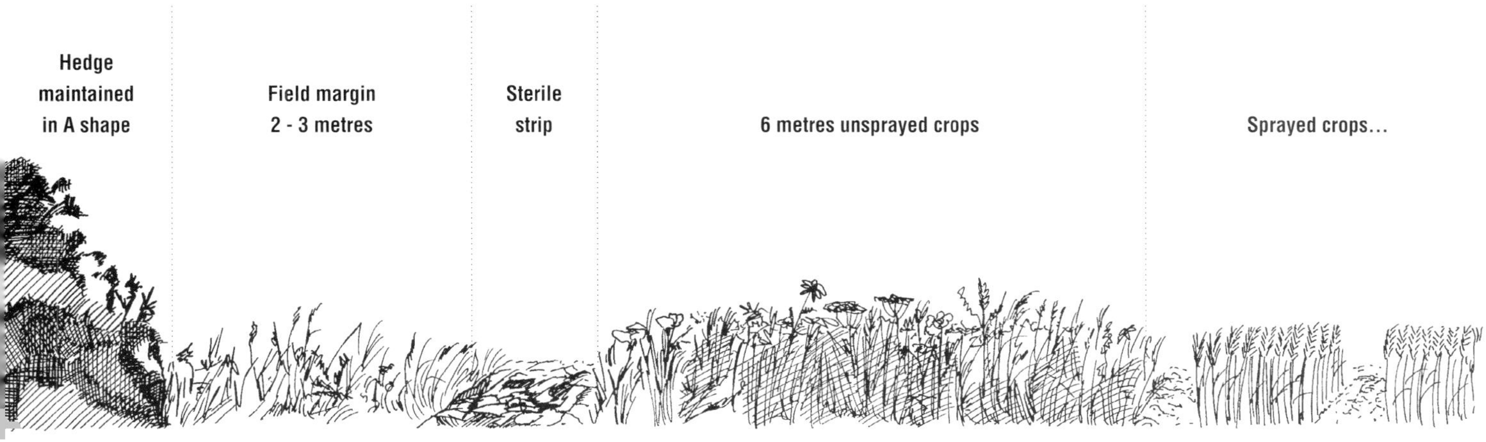

FORESTS AND WOODLANDS

Many butterflies will breed in woods and forests provided the right trees, plants and other features are present. In southern England up to forty species fly in woodland, although further north this number diminishes due to the cooler climate. For the truly woodland butterflies to flourish, relatively large areas of woods are necessary - probably several thousand acres but this may be partially interspersed by farmland or open country

The best butterfly woods are those with a predominance of deciduous trees, especially oak, and a complete age range from saplings to mature, dying and dead trees. As most butterflies feed and breed on ground flora, light must penetrate sufficiently for low-growing grasses, nectar flowers and larval foodplants to thrive. Surrounding trees give shelter and a warmer micro-climate than more exposed terrain. Thus, it will be seen that roads, rides, paths, glades, clearings or any similar breaks in the tree growth create a 'wood-edge' attractive to butterflies. Only the Purple Emperor and Purple Hairstreak are in a true sense woodland species as they will live and feed on the tree canopy only occasionally descending into open spaces.

An open woodland path with partial shade – a favoured habitat of the Speckled Wood and White Admiral.

In plantations, the wood-edges along roads and rides are where butterflies will be found, provided these are wide enough for sufficient sunlight to penetrate and vegetation to grow without being overshadowed by dense tree growth. Thick, shady forests, where no light reaches the ground because of the leaf canopy, will not support butter-flies. For this reason, mature plantations of conifers or some broad-leafed trees, such as mature beech, are poor wildlife habitats.

A few butterflies survive in regularly coppiced or cut wood-land and will die out very quickly once tree growth passes a certain stage of maturity. As little as two percent of woodland is now maintained for coppic-ing. Butterflies such as the Heath Fritillary and the violet-feeding fritillaries are under continuous threat and in many districts only able to survive in a few isolated locations which are still coppiced.

THREATS TO WOODLAND

It goes without saying that the main threat to woodland but-terflies is the destruction of mature and uneven-aged wood-land which is not allowed to regenerate or be replanted. The wholesale loss of woods where agriculture or land development takes place will obviously result in the disappearance of its butterflies.

Neglect of existing wood-lands is also a serious threat if rides and clearings are allowed to shade over as the tree canopy matures. The best woods for butterflies are those where well-planned management is sustained on a long term basis with conservation in mind. Plantations of coniferous trees will at best, reduce the number of species to those that can tolerate the conditions along rides and round the wood-edges where there is other vegetation.

CONSERVING WOODLAND

Protecting and managing mature and ancient deciduous wood-land is essential for the survival for the true woodland species which live on the tree canopy. When timber is cut it should only be taken from a small section of the woodland area so that the butterflies can move and survive in the adjoining woods. If possible some standard trees should be left; and if re-planting has to be with coniferous trees then some deciduous trees should be left or planted to grow around the plantation edges and in particular along rides. The Purple Emperor favours oak for the canopy and this tree is also the foodplant of the Purple Hairstreak.

WOOD-EDGE MANAGEMENT

The principles outlined here are designed primarily for a typical woodland road or ride, but can easily be applied to other wood-edges.

The objective of managing any wood-edge for butterflies is to encourage a variety of plant life to attract as many species as possible. There must be a profusion of ground cover so that nectar flowers and caterpillar foodplants can grow. This is best achieved by a grading from the tall mature trees sloping downward to younger trees, shrubs, undergrowth, grass and bare soil. Occasional gaps in such vegetation can vary the aspect, creating a greater variety, especially if there are breaks in the tree and shrub belts. Some butterflies require degrees of shading and thus should be considered when preparing management plans.

A ride with broad scalloped edges allowing growth of flowers, shrubs and grasses will support many butterflies.

A forest clearing at a ride junction designed to increase the woodland butterfly population.

Shape and aspect

Straight and level wood-edges or tracks may provide less shelter and create wind tunnels, whereas meandering, indented or curved edges can give better shelter and create sun traps. A wood with a southerly slope receives more sunshine, so the north side of an east-west ride should be given preference for maintaining the

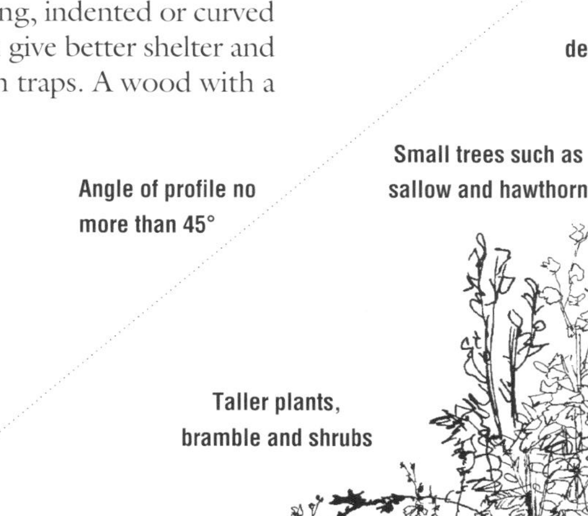

A woodland ride junction. The coniferous plantation in the background creates a sterile environment for butterflies whereas the foreground wood edges are cleared of timber to allow grasses, wildflowers and shrubs to grow in open sunlight. Mature deciduous standards remain along with some younger trees.

best conditions. A north-south wood-edge will tend to receive sunshine for the middle part of the day. Tree cutting should be along lines angled from the edge and at intervals to create small, south facing aspects for longer periods of sun. Such rides will be preferred by species such as the Speckled Wood, White Admiral and Wood White, which require partial shade.

A woodland 'panel' is cleared to benefit species such as the Heath Fritillary.

Road or ride width

For maximum sunshine, cut back trees to allow a width equivalent to the anticipated height of their eventual maturity. This applies on level ground, but sloping land will necessitate compensations, depending on angle and direction of slope.

Banks and ditches

In many woodlands, more recently constructed rides are flanked by drainage ditches and banks which help to maintain grasses and low vegetation. Steep edges provide useful bare areas of earth and a variety of soil conditions for many different plants. These should be cut and cleared on a rotational basis as for grassland.

Shaping wood-edges

The accompanying diagrams offer ideas on how to cut wood-edges in order to provide the best habitat conditions. Clearly these will have to be adapted to meet the prevailing features of the wood and if there is a special target species, then the particular requirements of that butterfly.

Bays or wedge-shaped clearances can be cut along wood-edges. Alternating or opposing bays may be placed along rides

At junctions, open clearings can be formed by cutting the corners to create 'box-junctions'. These can then be managed by coppicing or as grassland depending on the site priority.

Clearings and glades will only be useful if sufficient light can reach ground level for plants to grow. Glades may be formed naturally by groups of fallen trees. Artificial clearings need to be a minimum of thirty feet square to be of any use and probably larger where there are tall trees.

CUTTING PROGRAMME

The frequency of cutting will depend very much on the rate of growth of plants and whether the site is being managed for butterflies with special habitat requirements.

Grass tracks and paths

These should probably be cut annually if there is insufficient pedestrian or vehicle traffic to control vegetation.

Verges and grass swards

Cut in rotation to create grassland mosaic but not before late September. (See section on Grassland for further details).

Shrubs and undergrowth

Cut in rotation over a period of three to seven years, depending on speed of growth.

Trees

Unwanted trees, especially non-native species unsuitable for butterflies, should be removed every five to ten years, and only those which are to be preserved marked and left. A succession of the native trees should be allowed to grow, preferably in stands, so that there is a sequence of ages to provide variety. In the early stages of regeneration it may be necessary to protect saplings and young trees from deer and other animals.

COPPICE MANAGEMENT

Coppicing or regular clearance of woodland is essential for a number of butterflies. All require unshaded ground for the larval foodplants to grow and a warm position for the caterpillars to mature. Coppicing used to be widespread across the country but declined rapidly during the last century with an adverse effect not only on butterflies but on other wildlife as well.

The practice of coppicing probably dates back several thousand years and consists of the periodic cutting of wood with the stumps being left to re-grow for a further cropping. A wood is divided up into sections or 'panels' which are cut on a rotation of 7-35 years depending on the type of trees and thickness of wood required.

Butterfly colonies will survive on relatively small coppice panels but the optimum size is considered to be 5,000 to 10,000 square metres. The coppice panels should be planned in irregular shaped blocks and it is thought better to avoid straight edges or long narrow shapes. Ideally panels should be joined by open paths or rides which are regularly cleared to form corridor links especially for more sedentary species.

A south - facing sheltered wood edge – ideal conditions for many species, including the Purple Hairstreak.

Untended coppice soon overshadows ground cover and inhibits other vegetation.

Coppiced woodland will support the highest number of butterflies between the second and fifth years of regeneration, after which the growth becomes too dominant and conditions increasingly unacceptable for breeding purposes. Depending on the size of the coppiced woodland there should be a

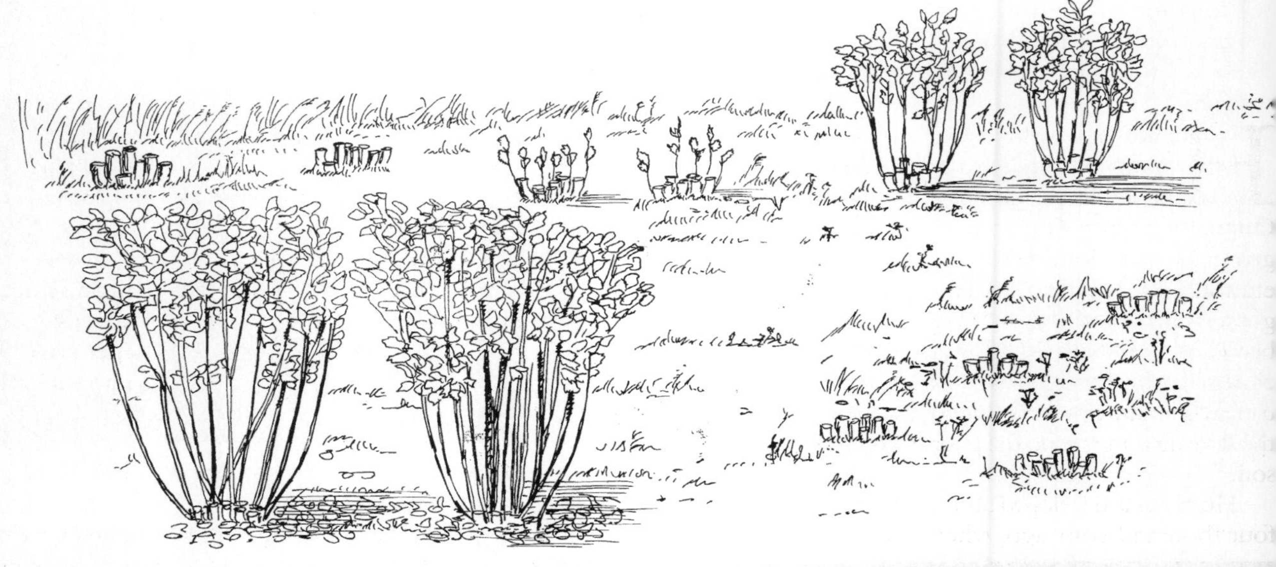

On the left, recent coppicing leaves ground unshaded allowing wildflowers, grasses and other 'soft' vegetation to grow. After a few years, new tree growth shades out sunlight thereby restricting ground cover until few, if any, plants survive. No butterflies can breed here.

substantial number of panels of 0-5 years growth in reasonable proximity. A maximum of 20% canopy cover should be allowed and woodland with a short coppice cycle of, say, 7 years will support the most plentiful colonies. Once tree growth starts to shade out ground cover and dominate, the butterfly nectar flowers and foodplants will disappear and it will be time to coppice again. This management must be continuous to ensure that the butterflies always have somewhere to move to.

Coppicing should only be undertaken with expert advice. Make sure that there is adequate labour for a long-term plan and, if possible, an outlet for the coppice products.

Typical woodland butterflies

- Lightly-shaded rides or glades (10-40% cover): Ringlet and Wood White
- Shaded rides or glades (40-90% cover): Speckled Wood and Green-veined White
- Newly cut woodland (including rides and coppice): Pearl-bordered, High Brown and Heath Fritillaries
- Dappled shade within woods or along edges: White Admiral and Silver-washed Fritillary
- Tree or shrub feeders on canopy: Purple Emperor and Purple, Black and White-letter Hairstreaks

FURTHER READING

- *Coppiced Woodlands: Their Management for Wildlife* (English Nature, 1990)
- *Woodland Butterflies* (Butterfly Conservation, 1989)
- *Woodlands - a Practical Handbook* (BTCV, 1988)

A Silver-washed Fritillary alights on brambles in a sunny clearing.

Woodland action summary

- Oppose felling and clearing for farming or development
- Protect deciduous broad-leaved woodland, especially old oak
- Discourage non-native species
- Oppose conversion to coniferous plantations
- Maintain rides and clearings for grasses, flowers and shrubs
- Ensure some sunlight reaches forest floor
- Manage wood-edges with variation, including graded vegetation
- Encourage continuity of traditional coppicing methods
- Seek habitat improvement as part of commercial management practices

HEATHLANDS FOR BUTTERFLIES

Heaths are wild, open areas of land, with poor, sandy acidic soils. Only a few types of grass will grow and the dominant vegetation is often heather and gorse. Pine and birch may become established, singly or in clumps, but reforestation is usually a lengthy process due to the slow growth rate on the poor soil.

Heathlands developed about four thousand years ago, when trees were cleared for grazing and crop growing. As most of these areas were on sandy soils, they readily lost nutrients and only a limited flora could survive.

It is estimated that between eighty and ninety percent of heathland has been lost during the last two hundred and fifty years. That which survives is scattered and fragmented, often leaving small and isolated habitats for heathland butterflies such as the Silver-studded Blue and Grayling.

Traditionally heathland was managed by fire - but now fragments are too small and too heavily used for this to be practical.

CONSERVATION

The main management objective is to maintain heather and natural grassland in large open areas and control invasion of bracken, trees or other vegetation from becoming dominant. This can be achieved by grazing, cutting or controlled burning. A heath should consist of heather of varying ages, ranging from pioneer, building, through to mature and regenerate, so that there is a constant variety. If such a pattern does not exist naturally, then it may be necessary to formulate a rotation of sections which are cleared over a cycle of ten to fifteen years, depending on growth rate.

Clearance may be by rotational block burning, but this must be carefully controlled and only undertaken where the safety of surrounding areas can be guaranteed. Cutting by drum mowers, flail or forage harvester

Pine and birch scrub is kept at bay on Hankley Common in Surrey. Patches of old heather are cleared to allow for regeneration.

Threats to heathland

- Ploughing and planting for regular agriculture and commercial forestry causes complete destruction of heathland
- Building development
- Invasion by pine, birch and bracken
- Recreational activities such as horse riding, motor cycling and large numbers of walkers may damage vegetation, leaving exposed soil which is easily eroded
- Fire starts easily during dry weather. Especially serious if frequent or over a large portion of a heath.

Typical heathland butterflies

Silver-studded Blue (amongst areas of young heather), Grayling (amongst grasses and areas of bare ground), Small Heath, Gatekeeper, Green Hairstreak and Small Copper. Many other grassland and migrant species may be found on heaths in small numbers.

Above: the Grayling butterfly, a typical heathland species.
Left: uncontrolled bracken invasion can obliterate all other vegetation from wide areas.

is safer and can be carried out during almost any weather. In some areas, old heather should be left to die down, thereby leaving space for new plants to regenerate.

Young saplings can be pulled up, but established trees will have to be cleared. Pine, if cut down , will not re-grow but birch will have to be dug out or chemically treated after cutting to prevent new shoots.

Bracken often invades heathland and should be controlled to protect a heathland habitat.

Grazing by cattle, ponies or sheep on heathland may be beneficial provided it is carefully controlled to avoid overgrazing and trampling which quickly kills heather. Similar principles will apply here as outlined in the section on grassland management.

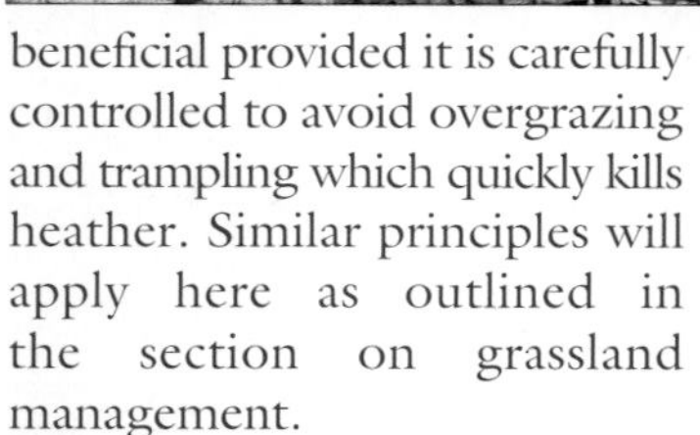

Sandy paths and tracks through heathland can easily be damaged by vehicles, horses and even an excessive number of walkers. Steeper slopes are extremely vulnerable and rainwater channelled along bare paths will wash away topsoil to form deep gulleys. The use of rights of way over heathland must be very carefully monitored. Where soil erosion takes place, access should be temporarily restricted and diverted. Often fencing is needed to guard soil protection measures and planting for regeneration.

It should be remembered that many rare birds and insects rely on heathland habitats. Management plans should take care to accommodate all important wildlife by preserving the necessary features for their survival.

Marginal heath areas are often important to butterflies as their richer plant life will attract more species.

Heathland action summary

- **Oppose destruction of heaths for farming, tree planting or land development**
- **Prevent invasion of trees, saplings, bracken or other degeneration**
- **Control recreational activities or direct to the least sensitive areas**
- **Prevent accidental fire**
- **Manage a rotation of heather growth**
- **Protect soil from undue erosion.**

Lowland heath in Southern England

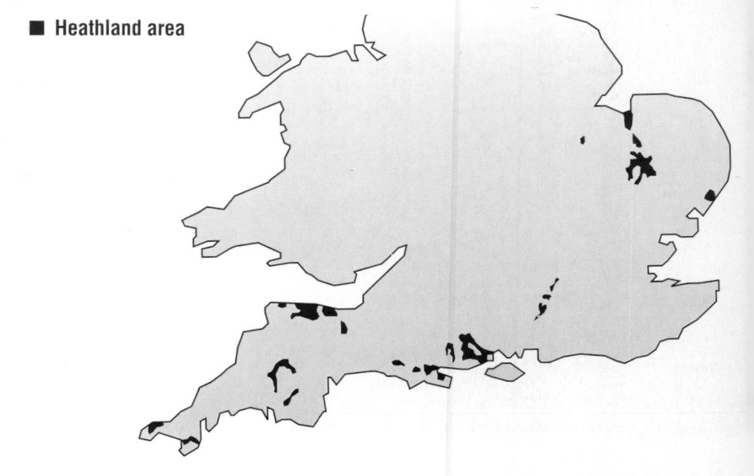

Right: in 1750 there were approx. 40,000 hectares of heath, by 1950 just 10,000 hectares were left. Today less than 6,000 hectares remain as shown here.

MOORLANDS AND MOUNTAINS

Moorlands and mountains in the British Isles do not support many species of butterfly due to the adverse climate and sparse vegetation. Nevertheless, there are several butterflies adapted to these conditions and in particular two species, the Mountain Ringlet and Scotch Argus, which were stranded in upland regions at the end of the last Ice Age.

Moorlands extend across many parts of the country, especially in central Wales, the Pennines and Scotland. Other areas include Dartmoor, Exmoor, the Cheviots and the North Yorkshire Moors. These are 'man-made' due to forest clearance on the higher ground by a gradual process over many hundreds of years or may never have supported trees.

Now moorlands are rough uncultivated areas with poor soil supporting only short vegetation such as grasses, heather, bracken and scrub.

CONSERVATION

Many of the important butterfly habitats in moorland and mountain regions are located in relatively isolated and undisturbed places.

Traditional methods of farming and land management need to be encouraged. The use of land for the production of a single crop whether it is arable or timber should be avoided.

The draining of moorlands should be discouraged as it will alter the vegetation and is not cost effective for agriculture.

The grazing of sheep and cattle should not be intensive in order to avoid a reduction in the variety of plant life and possible soil erosion.

The threats posed by land development and recreational activities need to be monitored very carefully whenever they are likely to damage or destroy butterfly habitats.

Near Bow Fell in the Lake District – typical habitat of the Mountain Ringlet.

Threats to moorland

- Afforestation by planting of coniferous trees which has destroyed nearly 10 million acres of moorland during the last 70 years
- Incorrect grazing levels - i.e. over- and under-grazing
- Improvement of more fertile moorland edges for agriculture: over-grazing and burning which cause deterioration and loss of the vegetation
- Recreational activities and developments such as reservoir and hydroelectric installations destroying moorland habitats

Typical moorland and mountain butterflies

Mountain Ringlet, Scotch Argus, Large Heath, Small Heath, Marsh Fritillary and Northern Brown Argus. Many other butterflies will be found on moorland edges where there is a richer and more varied vegetation.

Moorland action summary

- Oppose development of moorlands for intensive agriculture, tree-planting or land development
- Encourage traditional methods of agriculture and land management including light grazing
- Control recreational activities or direct to less sensitive areas
- Protect soil from erosion.

WETLANDS: PEAT BOGS, FENS & MARSHES

Peat extraction; alternatives are now available at garden centres.

Typical Swallowtail habitat in Norfolk.

Threats to wetlands

- Drainage for agriculture or building development purposes
- Changes in water table levels
- Water extraction from adjacent land
- Lack of management and cessation of traditional harvesting methods
- Chemical pollution of water
- Commercial peat extraction

Typical wetland butterflies

Peat bogs: Large Heath
Fens: Swallowtail
Marshes: Marsh Fritillary

Wetlands action summary

- Maintain existing water levels
- Oppose water extraction schemes which affect adjacent land
- Maintain and repair existing natural and artificial drainage systems
- Harvest reed beds traditionally
- Avoid chemical seepage from adjoining land
- Oppose commercial peat extraction

Wetlands are unique habitats which are host to some of our most vulnerable wildlife.

PEAT BOGS

The Large Heath is the only butterfly with the particular habitat requirement of peat bogs although some others may be found in such areas. This is often the case where peat bogs occur on heathlands, moorlands or in mountain regions.

Peat bogs are created by the constant waterlogging of land whereby the decomposition of vegetation is prevented and allowed to gradually accumulate over many years.

Traditionally peat was cut locally on a small scale for use as fuel but the greatest threat now is commercial cutting of peat moss to improve garden soil.

Many peat bogs are under threat from this commercial exploitation and must be protected. The southernmost habitat of the Large Heath butterfly was once threatened by a company wishing to extract peat on a far greater scale than ever before. Fortunately this site has now been purchased by English Nature and will be protected permanently.

In addition, the same type of activities which threaten heathlands and moorlands also affect peat bogs and their gradual destruction is of grave concern to conservationists. Over the last 100 years vast areas of peatland have also been lost to forestry and agriculture.

FENS AND MARSHES

The only remaining native species exclusively found in fens and marshes is the Swallowtail, now only resident in the Norfolk fenlands. The native Large Copper became extinct in the last century as a result of collecting, and the draining which lowered the water table in its stronghold fenland habitats. A continental subspecies has been reintroduced to a few sites with apparent success.

The habitat requirements of the Swallowtail have been widely researched and are now reasonably well understood. The main threat to its survival has been the cessation of traditional sedge cutting which then caused vegetation to grow rank thereby allowing the incursion of scrub.

Its main habitats are now centred in nature reserves but there is always the risk that these may be adversely affected by drainage or other farming activities in the vicinity.

Continuing vigilance is essential. A number of conservation organisations are closely monitoring the stronghold areas of the Swallowtail, so the more detailed aspects of its habitat conservation have not been included in this book.

COASTAL CLIFFS AND SAND DUNES

The many hundreds of miles of coastline around the British Isles offer a plentiful and often spectacular variety of butterfly habitats. Mostly these are grassland but scrub or woodland may grow in more sheltered or undisturbed places especially in the South. The grassland features of cliffs and sand dunes are attractive to many species of butterfly but there are none which have any specific requirement that is only found on the coast.

The Glanville Fritillary is only found along the south coast of the Isle of Wight among the rough under-cliff areas. However, this is due largely to the micro-climatic conditions there which enable the butterfly to survive at the northern edge of its range. The distribution of the Grayling and Dark Green Fritillary clearly illustrates the suitability of coastal habitats for this butterfly.

South coast undercliff on the Isle of Wight – the only habitat of the Glanville Fritillary.

A cliff path in Devon: the rich mix of habitats satisfies the needs of many native and migrant species.

The threats to all coastal areas are urban development, holiday and leisure activities, neglect and the encroachment of agriculture to the shoreline and cliff edges. Fortunately oil and chemical pollution to the seashore is not a serious threat to butterflies but it can be devastating for most other wildlife with which it comes into contact.

Damage to vegetation by vehicles and large numbers of trampling feet is a constant threat to sand dunes frequently causing serious erosion. Access should be limited and where necessary land protected for regeneration.

The National Trust through its Operation Neptune continues to acquire areas of natural coastline. It is encouraging, therefore, that, with careful management, many excellent butterfly habitats will be conserved.

Threats to coastal cliffs and sand dunes

- Urban development, caravan and camping sites, car parks
- Expansion of leisure activities
- Encroachment of agriculture to cliff edges and shorelines
- Walkers and holidaymakers trampling paths and vegetation
- Erosion by weather and the sea

Eroded sand dunes. Planting of Marram grass helps establish other vegetation and protect from walkers.

Coastline action summary

- Control land use and development
- Protect and expand shore margins from agricultural interference
- Limit vehicle access
- Contain and control leisure and recreational activity
- Maintain paths and tracks to prevent erosion
- Replant eroded areas and limit access until restored

URBAN AND SEMI-NATURAL HABITATS

Man has created an infinite variety of 'artificial' habitats, some of which are successfully adopted by the more common butterflies. Although these habitats are very different from the surrounding landscape, they are often oases of rich plant life ideal for butterflies.

Towns and suburbs

Within the urban environment, there are still areas where butterflies will find sufficiently favourable conditions to breed, especially in gardens, church-yards, cemeteries, parks and allotments.

Commerce and industry:

Factories, power stations, industrial estates, business parks, reservoirs, airfields, public utilities and the like often contain large areas of peripheral unused land.

Recreation and leisure

Sports grounds, golf courses, wildlife and theme parks, caravan and camping sites invariably have corners of spare land suitable for butterfly habitats.

Transport

Many miles of railway, motor-way, road and canal embank-ments and cuttings have become natural 'corridors' for wildlife.

Mines and quarries

Mineral extractions leave un-disturbed sheltered quarries and gravel pits which gradually regenerate a plant cover.

Redundant land

Rich wasteland has developed from land abandoned by industry, commerce, domestic refuse tips and agriculture.

CONSERVATION

Plans for the management of such areas should aim to create grassland with hedgerow margins, wood-edge and other habitat features. The general

Stave Hill Ecological Park in the London Docklands area, with the Canary Wharf tower looming in the background.

In this park limited grass cutting allows a diverse vegetation to flourish - and it saves money!

In central London this roundabout was once left wild, as seen here. It has now been planted with a formal arrangement of flowers quite unsuitable for butterflies.

principles for conservation will be the same and reference should be made to the appropriate sections. The scale may be smaller in built-up areas but it is surprising how many butterflies turn up in the urban environment given the right conditions. In one cemetery less than five miles from the centre of London over twenty species have been recorded.

The Living Churchyard information pack and audio-visual presentations are available from Church and Conservation Project, Arthur Rank Centre, Stoneleigh, Warwickshire CV8 2LZ.

This disused railway line in Norfolk has become an oasis of wildlife.

Urban and semi-natural habitats action summary

- Exploit land for local wildlife amenity use and encourage local involvement
- Leave open and rough wild areas and manage for butterflies
- Encourage growth of wildflowers and grasses
- Plant only native deciduous trees and shrubs
- Avoid excessive tidiness where vegetation can conveniently be left to grow
- Keep to a minimum all grass cutting or confine to essential areas - especially road verges
- Avoid the use of insecticides, herbicides and other chemicals
- Prevent rubbish tipping
- Prevent pollution by industrial or other waste
- Control access of people and vehicles which will unnecessarily disturb or damage habitat

Derelict urban and industrial land is soon invaded by buddleia, willow herb and nettlebeds. Other wildlife including butterflies may soon follow. Organisations like the London Wildlife Trust create reserves and oases of countryside in city centres.

GARDENS AND PARKS

Gardens can play an important role as a haven for butterflies - from the small 'patio' town garden to the extensive country house or estate garden.

Firstly, butterflies will be attracted to gardens by nectar-rich flowers. Secondly, a number of species may be encouraged to breed by leaving wild or unkempt areas for suitable plants, grasses and trees on which to lay eggs.

Sunny, sheltered spots should be chosen for butterfly nectar plants to flower in succession throughout spring, summer and autumn. Delightful combinations of colour and texture are easily grown in a 'butterfly' garden.

By leaving part of a lawn uncut wildflowers, herbs and grasses may be allowed to grow naturally. Most garden centres now sell wildflower seeds and plants - advice will be given as to the best species for particular soil conditions.

Shelter belts with privet (but not cut or trimmed), holly, hawthorn and buddleia have the added advantage of providing nectar flowers and also a refuge for hibernating Brimstone, Small Tortoiseshell, Comma and Peacock butterflies.

In orchards, grassland meadows between fruit trees can be managed for butterflies. Spraying should be avoided, or limited as much as possible. Windfalls should be left to rot so that autumn butterflies, such as the Red Admiral, can feed.

The use of insecticides and other chemicals in gardens should be avoided. Alternatives to peat, which are economic and at least as good, should be used. Leaf moulds, garden composts and commercial products will produce excellent results thereby conserving peat bog habitats.

NETTLE BEDS

Nettle beds will encourage the Small Tortoiseshell, Peacock, Comma and Red Admiral, the larvae of which all feed on this plant. Sheltered, sunny positions by walls, banks or hedges are preferred, although the Small

Typical garden butterflies

Large White, Small White (both often known as the Cabbage White), Small Tortoiseshell, Red Admiral, Peacock, Painted Lady, Comma and Holly Blue. These are the commonest garden butterflies even in town and city centres. In more rural areas plenty of others may be tempted to visit well-stocked butterfly gardens.

Peacocks and a Painted Lady feed on a buddleia bloom in a Devon garden.

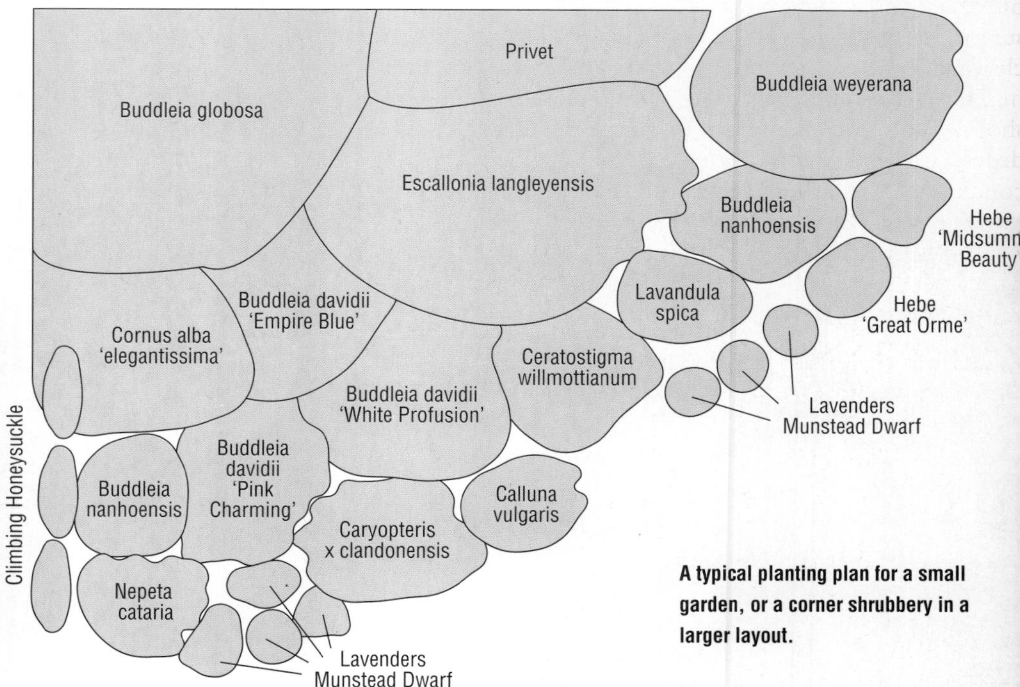

A typical planting plan for a small garden, or a corner shrubbery in a larger layout.

Uncut privet, good for nectar and shelter.

An open sunny position is sought by egg laying females, nettlebeds in shady corners are shunned. Grass cuttings encourage new growth to feed hungry caterpillars.

Tortoiseshell larvae are often found in more open spots. Females will choose early growth on which to lay their eggs. In summer it is worth cutting and clearing a one-third section of mature nettles to allow young shoots to reappear. Timing will depend on locality and weather conditions - in southern Britain the second half of June is probably best and further north perhaps a week or two later. Keep an eye out for feeding caterpillars and pupae and do not disturb them by cutting.

HIBERNATION

Most butterflies hibernate in the larval form, but a few as ova, pupa or adult butterfly. The Brimstone, Small Tortoiseshell, Comma, Peacock and Red Admiral hibernate as butterflies. Some times they may be found in houses or out-buildings during the winter. They should be removed gently from any heated or warm area to prevent activity and placed somewhere cool, dry and sheltered from which they can escape during a milder spell or with the onset of warmer spring weather.

A spring rock garden planted with nectar flowers for butterflies.

FURTHER READING

- *Gardening for Butterflies* (Butterfly Conservation 1989) booklet is recommended for more detailed information and advice.
- *Garden Plants for Butterflies* (Matthew Oates)an extensive guide for what to plant in your garden.
- *Starting a Butterfly Garden* (BTCV 1990).

Garden plants and shrubs for butterflies

Alyssum (eg. A. maritimum, A. saxatile)
Aster (eg. Callistephus chinensis)
Aubretia (Aubretia deltoidea)
Buddleia (B. davidii and most others)
Candytuft (Iberis, genus of 30 spp.)
Catmint (Nepeta - genus of 250 spp.)
Coreopsis (genus of 120 spp.)
Feverfew (Chrysanthemum parthenium)
Goldenrod (Solidago genus of 100 spp.)
Hebe (Hebe genus of 100 spp., formerly in genus Veronica, often still named so)
Honesty (Lunaria, genus of 3 spp.)
Hyssop (Hyssopus officianalis)
Ice Plant (Sedum spectabile only)
Lavender (Lavandula spica)
Marigold (Calendula officinalis, Tagetes genus of 50 spp. inc. T. erecta, T. patula)
Michaelmas Daisy (Aster, genus of 500 spp. - not to be confused with above)
Nasturtium (Tropaeolum majus)
Phlox (Phlox, genus of 66spp.)
Privet (Ligustrum spp. eg. ovalifolium)
Red Valerian (Centranthus ruber)
Scabious (Scabiosa, genus of 100 spp.)
Sweet William (Dianthus barbatus)
Verbena (Verbena, genus of 100 spp.)
Wallflower (Cheiranthus, 10 spp.)
Zinnia (Zinnia, genus of 20 spp.)

Night-scented plants for moths

Bladder Campion (Silene vulgaris)
Evening Primrose (Oenothera spp. inc. biennis, erythrosepala, parviflora, stricta)
Honeysuckle (Lonicera periclymenum)
Night-scented Stocks (Matthiola bicornis)
Petunia (Petunia, genus of 40 spp.)
Sweet Rocket (Hesperis matronalis)
Tobacco Plant (Nicotiana genus)
White Jasmine (Jasminum officinale)

CREATING WILDFLOWER MEADOWS

One of the most positive and rewarding aspects of butterfly conservation is the re-creation of wildflower and grassland meadows for butterflies. Heavily cropped arable fields or overgrazed pasture are butterfly deserts. Few butterflies will be seen and none can breed.

This can all be changed with careful planning, hard work and some money. An area of as little as an acre (but preferably more) may be transformed into a thriving habitat within a few years for at least 15 to 20 of our commoner butterflies, maybe more depending on the location and local conditions. There is no greater excitement than to see butterflies flying over flower rich meadows, where previously the land was inhospitable and there were none.

Many species will discover a suitable new habitat without encouragement. It is surprising how often and how quickly butterflies find sites, even those isolated by intensive farming or urban development. Localised or sedentary butterflies are less likely to be attracted.

Here are some ideas on the basics of re-establishing grassland habitats from previously farmed arable or pasture land.

THE SITE

Identify the most suitable area of land that can realistically be converted. A south-facing aspect with good shelter is ideal (see section on butterfly ecology). Less than an acre is unlikely to produce the best results.

SITE HISTORY

Check the history of the previous management and what fertilisers, pesticides or other chemicals have been applied. Soil tests may help to reveal any particular problems, such as heavy nitration which is unsuitable for most wildflowers and grasses.

THE PLAN

Plan the layout and landscaping to give a mosaic pattern of nectar and larval foodplants. Allow for dry and wet areas with a pond or stream if this is possible. Include means of access for both people and vehicles, paths, shelter belts for shrubs, and trees. Fencing will help to prevent damage by wild animals such as rabbits and deer.

THE SOIL

Herbicide spraying may be necessary to destroy unwanted weeds and grasses and may involve leaving the land fallow for a season. Many small seeds will not grow from large lumps

Five years ago this wildflower meadow in Norfolk was nothing more than a heavily nitrated arable field which was unsuitable for many foodplants and had no butterflies. Now an average of over twenty different species of butterfly breed on the site or are seen here every year. Careful management and a flexible cutting regime are essential to maintaining the diversity of plantlife.

Many tons of crushed chalk form the base of this breathtaking wildflower garden.

of soil so the surface should be broken up and levelled to a fine tilth.

SEEDS AND PLANTS

Select a variety of seed mixes from an expert supplier of locally harvested stock. Mixtures with 20 to 30 different plants are usual. It is best to choose native species which are easily germinated and quick growing. Study the lists of nectar and caterpillar foodplants in this book.

A different mix of grasses will be needed for paths and other areas which are to be regularly mown short.

SOWING AND PLANTING

Sowing should be done by hand to get an even spread. This is a skilled job so it is worth finding an expert. Ask your seed merchant for advice. Timing is also crucial as many seeds need a period of cold before germinating, so autumn is usually the best time, but follow the sowing instructions with your mixture.

The planting of shrubs and growing plants should be avoided in the dry summer months. If pot-grown plants are used they should be given space to grow and not be stifled by the surrounding vegetation.

SITE MANAGEMENT

Unless regularly monitored and tended the site will quickly degenerate to scrub.

First year

A light cut every 4-6 weeks during the summer will establish the new swards. Removal of unwanted weeds such as dock, bramble and thistle is advisable. Paths should be cut short as often as necessary for access and to give the butterflies basking spots.

Second year

Swards should be well established and your management programme formulated (see section on grassland habitats). A light cut in mid-summer is recommended, followed by a closer cut in late September. Continue to mow paths regularly.

Third year

A trim may be necessary in early spring and then the meadow can be left until a hay crop is taken towards the end of September. If ground conditions permit, a further tidying cut may be made in October or even later.

This regime will be the basis of the management pattern for future years. Shelter belts and shrubs should be growing well. Any dead or unhealthy plants ought to be replaced quickly so their replacements can still compete for light with those planted earlier.

Prune as necessary and cut back any shrubs which threaten to overwhelm their neighbours. Always remove all cut material especially dead grass which tends to mat and smother new growth.

RECORDING

From the outset record what is done on the site and then the progress from year to year. Retain your original plans and planting details. Photograph the land before starting and take a new set of pictures each year as it matures. You will soon have a fascinating visual history of change and development of the site.

Most importantly, record the butterflies seen on the land from the earliest opportunity (see the section *Putting butterflies on the map*). A methodical monitoring of all species will be your indicator of the success both of the creation and the management of the new butterfly habitat. Of equal interest will be the appearance of other wildlife, especially wild flowers which you

This old Sussex garden has been left wild to form a miniature nature reserve.

FOODPLANTS

WILDFLOWERS FOR NECTAR

All species of butterflies have to feed and most will go to a wide variety of flowers to take nectar. There are a few rarely seen on flowers as they prefer moisture, honey-dew, rotting fruit or other sources of nourishment.

The choice of nectar flowers changes during the season as they mature and bloom. Certain flowers, such as roses, are ignored by butterflies so, for conservation purposes, it is best to encourage the plants more popular to them. The list of wildflowers includes common species which will grow in suitable habitats throughout the British Isles.

A Brimstone feeds on a thistle head. Adult butterflies feed through their proboscis which is extended into flowers to extract nectar. Some species have a much shorter proboscis than others, so the types of flower they can feed from are more restricted.

A wildflower meadow – a variety of flowers available throughout the season.

Wild flowers as nectar sources for butterflies

Plant	*Months in flower*											
	J	F	M	A	M	J	J	A	S	O	N	D
Sallows	○	○	●	●	○	○	○	○	○	○	○	○
Coltsfoot	○	○	●	●	●	○	○	○	○	○	○	○
Primrose	○	○	●	●	●	●	○	○	○	○	○	○
Dandelion	○	○	●	●	●	●	●	○	○	○	○	○
Bluebell	○	○	○	●	●	●	○	○	○	○	○	○
Bugle	○	○	○	●	●	●	○	○	○	○	○	○
Cuckoo-Flower	○	○	○	●	●	●	○	○	○	○	○	○
Broom	○	○	○	○	●	●	○	○	○	○	○	○
Ragged Robin	○	○	○	○	●	●	●	○	○	○	○	○
Ox-eye Daisy	○	○	○	○	●	●	●	●	○	○	○	○
Red Campion	○	○	○	○	●	●	●	●	●	○	○	○
Clovers	○	○	○	○	●	●	●	●	●	○	○	○
Bramble	○	○	○	○	●	●	●	●	●	○	○	○
Violets	○	○	○	○	●	●	●	●	●	○	○	○
Bird's Foot Trefoil	○	○	○	○	●	●	●	●	●	○	○	○
Privet	○	○	○	○	○	●	●	○	○	○	○	○
Valerian	○	○	○	○	○	●	●	●	○	○	○	○
Meadowsweet	○	○	○	○	○	●	●	●	○	○	○	○
Sowthistle	○	○	○	○	○	●	●	●	○	○	○	○
Betony	○	○	○	○	○	●	●	●	●	○	○	○
Tormentil	○	○	○	○	○	●	●	●	●	○	○	○
Buddleias	○	○	○	○	○	●	●	●	●	●	○	○
Ragwort	○	○	○	○	○	●	●	●	●	●	○	○
Devil's Bit Scabious	○	○	○	○	○	●	●	●	●	●	○	○
Wild Thyme	○	○	○	○	○	○	●	●	○	○	○	○
Teasel	○	○	○	○	○	○	●	●	○	○	○	○
Hemp Agrimony	○	○	○	○	○	○	●	●	●	○	○	○
Knapweeds	○	○	○	○	○	○	●	●	●	○	○	○
Heathers	○	○	○	○	○	○	●	●	●	○	○	○
Marjoram	○	○	○	○	○	○	●	●	●	○	○	○
Watermint	○	○	○	○	○	○	●	●	●	●	○	○
Thistles	○	○	○	○	○	○	●	●	●	●	○	○
Fleabane	○	○	○	○	○	○	○	●	●	○	○	○
Ivy	○	○	○	○	○	○	○	○	●	●	○	○

CATERPILLAR FOODPLANTS

Butterflies breed only on sites where the preferred caterpillar foodplants ;row. These must be growing n an acceptable position for egg aying and at a time when the :aterpillars feed. The caterpillars)f some species will eat a variety)f plants but most are highly elective, and a few will eat only)ne type of vegetation. The .bility to adapt to a new food-)lant is thought to be the key to he survival and resurgence of ome species, such as the Comma. In compiling a list of :aterpillar foodplants it has to be :mphasised that regional and ocal variations are frequent,)articularly with grass feeding pecies.

Seasonal, unusual or extreme veather conditions may also .ffect the choice of foodplant ind feeding habits. Drought or 'ery hot temperatures may vither and even kill smaller)lants and grasses, resulting in :gg laying on alternative but not iecessarily suitable plants. Caterpillars may be forced to eek other vegetation although . change of diet does not ;uarantee survival. In captivity nany species will feed on plants vhich they ignore in the wild.

The foodplant list overleaf herefore include the most :ommonly used but is by no neans exclusive.

A Brimstone caterpillar resting on a partially eaten Buckthorn leaf.

Cowslips, the foodplant of the Duke of Burgundy.

A Small Skipper laying (ovipositing) on a grass stem.

A Small Tortoiseshell caterpillar suns itself on stinging nettles.

British butterflies and their caterpillar foodplants

Chequered Skipper
Purple Moor-grass and other grasses (Scotland). False Brome (English colonies).

Small Skipper
Yorkshire Fog, Creeping Soft- and False brome grasses.

Essex Skipper
Cock's Foot and Creeping Soft-grass.

Lulworth Skipper
Tor-grass

Silver-spotted Skipper
Sheep's Fescue grass (Eggs may be laid on adjacent plants)

Large Skipper
Cock's Foot grass and some other grasses possibly depending on location.

Dingy Skipper
Bird's-foot Trefoil or Great Bird's-foot Trefoil and Horse-shoe Vetch in some localities.

Grizzled Skipper
Wild Strawberry, Bramble and other potentillas. Raspberry and Agrimony in captivity.

Swallowtail
Milk-parsley. Fennel, and Carrot in captivity.

Wood White
Meadow Vetchling, Bitter Vetch, trefoils and some other legumes.

Brimstone
Buckthorn and Alder Buckthorn.

Large White
Crucifers. Cultivated or wild brassicas, especially cabbage and garden nasturtium.

Small White
Similar to Large White.

Green-veined White
Lady's Smock or Cuckoo-flower, Garlic Mustard, Hedge Mustard and other crucifers.

Orange Tip
Similar to Green-veined White. Will usually eat the flower heads and seed pods. Also Garden Honesty.

Green Hairstreak
Common Rock-rose, Gorse, Broom, a variety of legumes and shrub flowers.

Brown Hairstreak
Blackthorn and most other prunii.

Purple Hairstreak
Most Oaks.

White-letter Hairstreak
Most Elm with preference for Wych Elm.

Black Hairstreak
Blackthorn and wild plums.

Small Copper
Common and Sheep's Sorrel and Dock spp.

Small Blue
Kidney Vetch (flower heads)

Silver-studded Blue
Flowers and young shoots of heathers, heaths and gorses on heaths. Rock-rose and Bird's-foot Trefoil on grasslands.

Brown Argus
Rock-rose or, if unavailable, Common Stork's-bill and Dove's-foot Cranesbill.

Northern Brown Argus
Common Rock-rose.

Common Blue
Bird's-foot Trefoil, other trefoils and variety of other plants.

Chalkhill Blue
Horseshoe Vetch principally but also other vetches, trefoils and variety of plants.

Adonis Blue
Horseshoe Vetch.

Holly Blue
Holly flower buds (spring brood). Ivy flower buds (autumn brood). Also other shrubs.

Duke of Burgundy
Cowslip, occasionally Primrose, and in captivity primulas.

White Admiral
Honeysuckle.

Purple Emperor
Common Sallow and Crack Willow.

Red Admiral
Common stinging Nettle and Small Nettle.

Small Tortoiseshell
Stinging Nettle and Small Nettle.

Peacock
Stinging Nettle, possibly Hop.

Comma
Stinging Nettle, occasionally Hop, Currant, Elm or Sallow.

Small Pearl-bordered Fritillary
Common Dog Violet, in damper places Marsh Violet. Other violets.

Pearl-bordered Fritillary
Common Dog Violet and other violets.

High Brown Fritillary
Common Dog Violet and other violets.

Dark Green Fritillary
Common Dog Violet in wood-land, Hairy Violet in downland, Marsh Violet in other locations.

Silver-washed Fritillary
Common Dog Violet and probably other violets as well.

Marsh Fritillary
Devil's-bit Scabious. Honey-suckle, Snowberry and Teasel in captivity.

Glanville Fritillary
Ribwort Plantain and if necessary Bucks-horn Plantain.

Heath Fritillary
Common Cow-wheat in woodland and heathland, Rib-wort Plantain and Germander Speedwell in grassland and some other plants.

Speckled Wood
Many grasses including False Brome, Cock's Foot and Yorkshire Fog.

Wall Brown
Grasses including Tor-, Purple Moor- and Cock's Foot grass.

Mountain Ringlet
Mat grass.

Scotch Argus
Purple or Blue moor grass primarily but possibly fescues and some other grasses.

Marbled White
Sheep's Fescue, Red Fescue and possibly other related grasses. Tor-grass and Upright Brome important on chalk.

Grayling
Bristle Bent, Sheep's Fescue, Early Hair- and other grasses depending on location. Also broad leaf grasses in captivity.

Gatekeeper
Narrow blade grasses including fescues and bents.

Meadow Brown
Narrower blade meadow grasses including fescues and bents.

Small Heath
Narrow blade grasses including fescues and bents.

Large Heath
Hare's-tail Cottongrass, White-beaked Sedge and some other grasses.

Ringlet
Tufted Hair-grass and many other grasses specially in captivity.

MIGRANTS

Bath White
Wild crucifers.

Clouded Yellow
Clover, Lucerne and Bird's-Foot trefoil.

Pale Clouded Yellow
Trefoils and clovers.

Berger's Clouded Yellow
Trefoils and clovers.

Long-tailed Blue
Ever-lasting Pea, Bladder Senna and other legume species.

Large Tortoiseshell
Elm species, occasionally Sallow, Willow, Cherry and some other trees or shrubs.

Painted Lady
Spear Thistle and Marsh Thistle preferred, also Creeping Thistle and Stinging Nettle.

Camberwell Beauty
Willows and sallows.

Queen of Spain Fritillary
Species of violets, Lucerne and Borage. Pansy in captivity.

Monarch
Milkweeds.

EXTINCT SPECIES

Black-veined White
Blackthorn, Hawthorn and some other wild or cultivated fruit trees.

Large Copper*
Water Dock.

Mazarine Blue
Red Clover. Kidney Vetch in captivity.

Large Blue*
Wild Thyme (presence of the ant species *Myrmica sabuleti* necessary on site).

* *re-introduced using continental subspecies*

CARING FOR HABITATS

FIGHTING LAND DEVELOPMENT

What can be done to protect a good butterfly habitat or area? Watch out for any indication of potential threats. Development may be preceded by land sales - look for estate agents' boards and property advertisements.

No development of land or change of use should take place without planning permission. Monitor planning applications to local councils. There can be two main stages, the first being application for 'outline' and followed by 'detailed consent'. The former is the most important as it establishes the principle, so it is the point at which objections must be lodged against unacceptable proposals. Alternatively, detailed consent may be sought immediately.

Planning law is highly complex, requiring expert advice. However, any member of the public may inspect a planning application lodged at the local Town Hall and planning officers will always explain the proposal.

Find out as much as you can about the ownership and any proposals for changes. Alert your local Butterfly Conservation branch, the local Naturalist or County Trust and other appropriate wildlife organisations. Tell them of possible dangers and enlist support.

A planned approach under the auspices of a conservation organisation should be made to the landowners or developers. Make known the exact dangers to butterflies with a view to preventing or at least minimising damage to the threatened habitat. Nowadays there is much greater awareness and sympathy for 'green' issues. Large companies and commercial organisations are increasingly sensitive and willing to co-operate if asked in the right way. Indeed, sponsorship and other resources are often available so the prime objective should be to work with and not against the developers - conflict is always the least productive approach.

At Aston Rowant the M40 cut through the middle of an important nature reserve. The more recently built M40 extension was rerouted to avoid Bernwood Forest, home of the Black Hairstreak.

Commercial sand and gravel extraction eats up another coastal heathland site.

NEXT STEPS

If negotiations fail and a campaign is needed to protect the habitat, this must be carefully researched and formulated to create the greatest impact and credibility:

- Consider publicity through press, TV and radio (local and national).
- Organise petitions, letter writing and public meetings.
- Contact or write to the local councillors and MPs, the Minister for the Environment, English Nature and other conservation organisations.
- Make representations to planning authorities, planning appeals and public inquiries.

To be successful a campaign of this nature needs careful co-ordination amongst all interested parties. It will be time-consuming and probably require financial backing.

NATURE RESERVES FOR BUTTERFLIES

Ranmore Common, near Dorking, is one of the finest downland butterfly habitats in the country. Active conservation work maintains this National Trust land which would otherwise become overgrown and unsuitable for many species.

The majority of the best butterfly habitats is now found on nature reserves or protected land. These areas have exceptional characteristics to support rare and diverse wildlife, so are not necessarily conserved for butterflies alone. However, success in maintaining butterfly populations on some reserves has been mixed as protection is not enough and positive management is also needed.

As mentioned under the section on Butterflies and the Law, reserves owned by official organisations have strict rules to protect wildlife. Undoubtedly butterflies are safer with this protection, but enforcement by wardens is labour intensive and there is still some illicit collecting. Voluntary helpers are often welcomed by the organisations responsible for protecting important sites with rare butterflies and official warden patrols do act as a deterrent.

The same principles of management and conservation apply to these habitats. It is crucial that conservationists establish close links and co-operation with the landowners. They are not always aware of the richness or value of their butterfly habitat. On the other hand, the larger land-owning organisations have conservation departments which should be contacted by anyone interested in their work.

Nevertheless, there are still many butterfly sites needing safe protection and the following section outlines the steps necessary to set up a butterfly reserve.

CREATING RESERVES FOR BUTTERFLIES

Conservationists frequently find it necessary to gain a controlling interest in land where the future safety of its butterflies cannot otherwise be guaranteed. The objective is to ensure the land is sympathetically managed for whatever specifically targeted butterflies are resident there or are to be established. A management plan is essential and in order to establish the broader issues reference should be made to *Management Plans - A Guide to their Preparation and Use* (Countryside Commission, 1986).

In a sense, the establishment of a reserve is a last resort where there is no other chance of the site remaining as a viable habitat. Very often the most effective course is to work in conjunction with other conservation groups as joint knowledge and resources may well be needed. This is a brief outline of the main steps for setting up a reserve:

The land

Identify the area of special interest and suitable boundaries of the proposed reserve. Make contact with owner to explain the purpose of interest. Such approaches should be made with great care as unsympathetic landowners have in the past destroyed habitats when alerted to a possible conservation priority hindering development. Obtain permission to enter on land if there are no suitable public rights of access.

The butterflies

Survey and monitor butterfly population to identify rare and vulnerable species and assess suitability of site for sustaining long-term permanent colonies of targeted species. Do not forget to look for the early life stages; eggs, caterpillars and pupae. This stage will require detailed expert advice, and should include a clear indication of the future management commitment and resources required. At the same time an assessment of other wildlife will indicate whether butterflies should be the priority for future management plans.

Management plan

A management plan should be prepared with expert input from entomologists, botanists and other naturalists preferably with knowledge of the locality. Problems may arise from activities on the adjoining land. It is important to anticipate anything that might interfere with the successful management of the reserve.

Resources and cost

List the resources needed to set up and maintain the reserve. Initially, fencing, notices and site clearance are likely to be expensive. Once established, the cost of management must be projected and a budget for the reserve agreed.

Tenure

Consider alternatives for setting up the reserve, such as the purchase of the freehold or a long leasehold, the granting of a lease or licence or a land management agreement. Rights of access to the land, rights of way of others, sporting rights, wayleaves and covenants affecting uses must be identified and clearly understood. The services of a conveyancing lawyer in the early stages of negotiation is likely to avoid tricky legal pitfalls later on.

Magdalen Hill Down Reserve, near Winchester in Hampshire, May 1992. Scrub clearance and fencing was a major operation carried out largely by volunteers.

Acquisition

A valuation of the land may be needed from a suitably qualified local estate agent or surveyor. (A charitable organisation buying land must show that the price is no more than the open market value - a nominal or below-market value price is obviously preferable).

Fund raising

The money needed must come either from existing funds or a fund-raising publicity campaign, sponsorship and/or grants.

WHO DOES THE WORK?

The acquisition of a reserve is a major operation and highly labour-intensive. Help will be needed from people with business expertise and professional qualifications, ideally on a voluntary unpaid basis but if not the cost must be included in the cost of establishing the reserve.

Volunteers to work on the land and plenty of enthusiastic and reliable administrative back-up are essential for the long-term success of the reserve. Establish contact with contractors specialising in conservation work which cannot be undertaken by volunteers. Find sympathetic farmers willing to graze land in accordance with an agreed programme or rota.

• *Organisation of a Local Conservation Group* (BTCV) is a book which comprehensively explains the whole process of setting up and running a successful conservation group.

Summary plan of action for conservation work

■ **SURVEY** Examine the site carefully and see what is already there. Look for butterflies, eggs, caterpillars, foodplants, nectar plants, roosting sites, breeding areas, sunny glades, shelter spots, etc. Examine and record other wildlife (plants, animals and insects) which live on the site.

■ **DEFINE OBJECTIVES** Decide what can be done to improve particular populations; to attract species breeding in the area; to create new features, provide shelter and clear overgrown areas. *Never* change conditions to the detriment of existing populations. Have a clear idea of what you want to achieve in (say) one year and after five years.

■ **PLAN THE WORK** Get expert advice. Estimate costs of fencing, tools, plants etc. Plan the timescale of the work and order of priorities. Find people to help. Read up on how to do things. Look at examples that have been successful in your area.

■ **CARRY OUT THE PLAN** Keep a record of what you do and when. Never change the whole of an area at once - do a section at a time. Plan the best time to do things. Think how your activities are affecting all the wildlife. In particular, avoid disturbing breeding birds and animals.

■ **MONITOR THE RESULTS** Does your work have the desired effect? Take 'before' and 'after' photographs. Monitor populations and compare treated and untreated areas. Are there any unexpected bonuses or problems? If so, do you need to revise your master plan?

Section Three
PROTECTION, EDUCATION AND RESEARCH

We now move on from the ecology of butterflies and the practicalities of on-site land management. Insects themselves often need protection, especially those which are a rarity or exceptionally beautiful. Sadly, a small minority of people may threaten butterflies in the wild by collecting or other activities, which means their survival can only be guaranteed through effective operation of the law.

Controlling the use and development of land is essential for the preservation of habitats. Until recent years legislation has rarely, if ever, been formulated with this objective in mind and it still has a long way to go.

However, attitudes are changing and the legal framework now evolving in this country reflects the growing awareness of the dangers to our natural heritage. The focus is the Wildlife and Countryside Act 1981 but its scope is limited, development slow and enforcement problematic to say the least. Bye-laws and regulations issued by land-owning bodies, such as the National Trust and Forestry Commission, offer a degree of protection; but they are geographically limited and only cover relatively small parts of the country.

How we behave in the countryside and treat our wildlife demands careful thought and reflection. For example, should butterfly collecting be allowed at all? If so, should it be a free-for-all or severely restricted? Another issue is the re-establishment of butterfly colonies; is it wise to transfer a species from one area or region to release it in another? What might be the implications of such action?

Laws, regulation and control tend to be negative and unpopular. Conservation is positive and, therefore, awareness and understanding must be gained from education and scientific research. This can start at an early age when ideas and attitudes are formed. Learning to observe, study and record is well within our grasp. The results will aid more complex scientific research in unlocking the secrets of butterfly life.

BUTTERFLIES AND THE LAW

There are two ways in which the Law can help butterflies: by protecting particular species and by protecting suitable habitats from destruction.

PROTECTED SPECIES

The Wildlife and Countryside Act 1981, Schedule 5, gives limited protection to four species of butterfly and five moths in the United Kingdom. These are shown in the box above. Under this Act it is a criminal offence if any person:

• intentionally kills, injures or takes a specimen from the wild (this, in effect, prohibits collecting for any purpose)

• has in their possession or control any live or dead wild specimen or any part of or anything derived from such specimen

• sells, offers or exposes for sale, or has in their possession or transports for the purpose of sale, any live or dead specimen (the objective is to prevent trading)

• publishes or causes to be published any advertisement likely to be understood as conveying that they buy or sell, or intend to buy or sell any of those things (this prohibits traders from advertising).

Butterflies and moths protected by Schedule 5

Swallowtail Butterfly
Large Blue
Heath Fritillary
High Brown Fritillary
Barberry Carpet Moth
Black veined Moth
Essex Emerald
New Forest Burnet
Reddish Buff

It is presumed in law that the specimen in question comes from the wild, unless the contrary is shown. Ova, larva and pupa as well as the mature insects are covered by the Act.

A strict interpretation of the Act also means that it applies to specimens of the same species of butterflies or moths caught not only in the UK but also Europe or elsewhere abroad (even non-British sub-species).

Penalties

The penalties on conviction currently include a fine of up to £2,000 per specimen, forfeiture of all specimens in respect of which a conviction is secured and possible forfeiture of any vehicle or other items used to commit the offence.

Licences

English Nature may grant a licence for collecting or possession which otherwise would be an offence. The Act provides for strict limitations and conditions which may be included in the licence. Licences may only be granted for certain purposes relating to scientific research, education, conservation or introductions to particular areas.

The Secretary of State (Department of the Environment) may also grant licences in respect of trading and advertising subject to the same conditions set out in the Act.

PARTIALLY PROTECTED BUTTERFLIES

Under the Wildlife and Countryside Act 1981 (Variation of Schedule) Order 1989, 21 species of butterflies received protection by the imposition of a ban on trading and advertising. This came into effect on 28 June 1989 and covers the species listed in the box below.

The trading or advertising for sale of all these butterflies is therefore illegal unless it is conducted in accordance with the terms of a licence granted by the Department of the Environment.

Thus, a wild origin egg cannot be traded nor the subsequent larva, pupa or adult without a licence. Bred stock must

Butterflies given partial protection by law

Northern Brown Argus	Duke of Burgundy Fritillary	Large Heath
Adonis Blue	Glanville Fritillary	Mountain Ringlet
Chalkhill Blue	Marsh Fritillary	Chequered Skipper
Silver-studded Blue	Pearl-bordered Fritillary	Lulworth Skipper
Small Blue	Black Hairstreak	Silver-spotted Skipper
Large Copper	Brown Hairstreak	Large Tortoiseshell
Purple Emperor	White-letter Hairstreak	Wood White

A European Swallowtail, similar in appearance to, but genetically very different from the English sub-species. It is illegal to release foreign butterflies in this country, even if the species are considered native.

The High Brown Fritillary, the most recent addition to the list of species given 'Schedule 5' protection by the Wildlife and Countryside Act.

derive from specimens caught wild before June 1989 if it is to be traded. Since the onus of proving origin lies with the seller he or she should inform the Department of the Environment to forestall an investigation.

Legislation needs tightening, so that all trading is under licence regardless of stock origin. The practice of obtaining female butterflies from the wild for trade breeding or strengthening stock genetics should be strongly discouraged. In certain circumstances there may be a case for taking such stock for breeding by experts, provided this is done under strict supervision. However, at present it is absolutely clear that it is illegal to trade in any of the 21 species caught in the wild after June 1989 or their progeny.

RELEASING FOREIGN INSECTS

The Act also prohibits the release of any insect which is not of a kind ordinarily resident in or a regular migrant visitor to Great Britain in a wild state. This means that anyone breeding butterflies or moths from stock which was caught abroad should not release these into the wild in this country.

The genetics of our resident species need safeguarding against adverse contamination by foreign stock, especially where sub-species or rare butterflies are involved. There will be genetic physiological differences even though the insects may appear to be identical.

PROTECTED LAND

Land owned or controlled by national or local organisations such as the National Trust, Forestry Commission, English Nature, County Trusts or conservation organisations, public utilities and local councils in theory, give a safe haven for butterflies.

In order to protect wildlife there are strict bye-laws, regulations or conditions relating to public access and behaviour. In most instances a breach of these rules will be a criminal offence. Collecting, even common species, is usually prohibited without an official written permit from the appropriate authority. Permits are usually granted only where the applicant is undertaking genuine scientific research.

Additional protection is afforded to lands designated as Sites of Special Scientific Interest (SSSI) where no collecting is allowed except with the consent of the landowner and English Nature.

Military training areas and ranges contain some of the best undisturbed countryside for all wildlife because they are subject to strict limitations on access.

STRENGTHENING THE LEGISLATION

There is now a strong body of opinion that full protection, i.e. collecting and possession, should be extended to most, if not all, of the 21 species partially covered by the Wildlife and Countryside Act 1981. If sufficient pressure could be brought to bear on government through the Department of the Environment, collecting and possession bans could be imposed by a simple Variation Order. Controls on taking live stock for breeding purposes under licence will be difficult to formulate and enforce.

- *Legislation to Conserve Insects in Europe* (Pamphlet No. 13. The Amateur Entomologists' Society 1987)

COLLECTING BUTTERFLIES

Few issues have seen such heated argument among entomologists and conservationists. Inadequate and often unenforced legislation adds to the confusion by implicitly condoning collection of species not covered by law.

USEFUL COLLECTIONS

In past decades, most collectors killed scores of butterflies just for the pleasure of filling their cabinets with beautiful displays. Often this was done with little scientific purpose in mind or understanding of the dangers of over-collecting or the ecological repercussions.

However, systematically arranged collections with full data (of where and when each specimen was caught) are a valuable legacy as many have been acquired by museums and academic institutes. Here they should be carefully preserved for entomologists and the public to study and carry out research.

The data relating to exact location and date of capture have assisted in building up records of distribution during the nineteenth century. In view of the wealth of material in collections and the extent of modern methods of recording live butterflies further collecting is, therefore, of little benefit to anyone.

It should also be recognised that the study and recording of the majority of other insects cannot rely on field identification and that collecting such specimens may be necessary for information on which to base conservation strategies.

A 19th century collection of Large Blues, now preserved in a museum.

OBJECTIONS TO COLLECTING

Strong moral and ethical objections to collecting are becoming more widespread as a result of the greater public awareness of conservation issues. Catching and killing butterflies just to make a collection serves no useful or scientific purpose and only reduces butterfly numbers. The unnecessary destruction of harmless creatures for personal gratification is morally indefensible and abhorrent to most people.

On the other hand, there are many people who started their interest in butterflies by collecting and many have since become very keen conservationists. As good butterfly sites become scarcer more people will want to see them and with decline of the butterfly population it is nowadays totally inappropriate for collecting to continue. Therefore, the acceptability of collecting will further diminish and be superseded by more constructive pastimes.

THE DANGERS OF COLLECTING

Whilst the worst threat to butterflies still remains habitat loss, the indiscriminate activity of some commercial traders and a minority of collectors are a threat. This is especially so in respect of the rarer species or where small or isolated colonies are plundered and in some known instances wiped out.

A total ban on collecting would undoubtedly alienate some sectors of the entomological world and perhaps force illicit trading underground. It is unlikely that the collecting of common species has resulted in any serious damage, but it is not worth the risk and should be discouraged strongly except for thoughtful scientific and educational purposes.

An additional problem is the enthusiasts who collect aberrations. The rarity of such variations necessitates the taking of vast numbers of the early life stages and therefore endangering colonies.

PROTECTING BUTTERFLIES

The JCCBI (Joint Council for the Conservation of British Insects) have issued a code of practice for insect collectors, but this represents only a minimal position held by scientists and some conservationists for insects in general. It forms a baseline from which further progress and development of attitudes may evolve if the code is strengthened.

BREEDING BUTTERFLIES IN CAPTIVITY

Breeding butterflies in captivity can serve a number of useful conservation purposes. It should be approached with caution, carefully planned and recorded.

A few common British species will breed very easily under artificial conditions. However, a considerable degree of skill and knowledge is needed to successfully rear most other species. It is even more difficult to keep them going from one generation to another. Techniques are now advancing enough for all the British butterflies to be bred from year to year when only recently this was thought to be impossible.

In the near future a 'conservation stock' of endangered species could be bred continuously in quantity for re-establishing colonies. If this had been possible for the Large Blue (extinct in 1979) the re-establishment of this butterfly might have been achieved with native rather than foreign livestock.

Ideally, competent breeders keeping stock should be encouraged to co-operate by exchanging livestock regularly to preserve a strong genetic pool. This type of breeding should be undertaken only by reasonably experienced entomologists with help from those who have already been successful.

Much knowledge can be gained by breeding butterflies although artificial conditions may not always truly reflect behaviour in the wild. The complete life-cycle can be observed in close detail and with patient recording much useful data assembled. An insight into the environmental requirements and behavioural patterns during the life-cycle greatly assists conservation work in natural habitats.

Breeding butterflies – some simple rules

- **Only take the commonest species from the wild. Make sure you avoid any species protected by the Wildlife and Countryside Act 1981.**
- **Avoid depleting wild colonies by collecting no more than a very few specimens (egg, larvae, or pupae) from the same site; egg bearing female butterflies should not be captured.**
- **Do not collect from nature reserves. Observe all local bye-laws.**
- **Make sure you have a plentiful supply of the correct foodplant for as long as the larvae will be feeding.**
- **Keep larvae in a ventilated container in the light but out of direct sunlight. If condensation appears, increase ventilation to prevent damp and mould growth.**
- **Provide fresh foodplant as often as needed. Clean the cage at the same time.**
- **Fix some twigs or sticks for the larvae to attach themselves to for pupation. Emerging butterflies will hang from the pupa case for their wings to expand and dry so allow plenty of space for fully grown insects.**
- **Release butterflies as soon as possible in locality of origin or suitable habitat nearby. Retain minimum number of stock if further breeding is planned.**
- **Over-wintering should be in outdoor conditions. Potted foodplants with hibernating larvae or pupae should not become waterlogged. Do not keep in warm room as this may induce unnaturally fast development or early emergence during the wrong season.**
- **Breeding of foreign or exotic species is not to be encouraged unless for a specific scientific or conservation purpose. Releasing such butterflies will result in death and in any event is against the law in the UK.**

There are many types of breeding cage for caterpillars - this cylindrical design is one of the simplest.

GETTING STARTED

For beginners it is suggested that larvae of the Small Tortoiseshell, Peacock, Large White or Small White are suitable and easily found. Many of the larger common moths, especially the Hawk-moths, will happily breed in captivity.

EDUCATION

Many butterfly enthusiasts enjoy the hobby of breeding common species and can gradually progress to the more exacting ones. This stimulates an interest in conservation and a better understanding of what butterflies need in order to survive.

GOOD READING

The Amateur Entomologist's Society's booklets *Breeding the British Butterflies* by P.W. Cribb and *Breeding the European Hawk-Moths* by P. Sokoloff are excellent guides. Most butterfly books have sections on simple breeding techniques. The only other book solely devoted to the topic is *Breeding the European Butterflies and Moths* by E. Friedrich, 1986 published by Harley Books.

RELEASING BUTTERFLIES IN THE WILD

The artificial release of butterflies is a potential means of establishing new colonies. Less mobile species can be moved to habitats which they would not otherwise exploit because they are too far away or isolated.

This method of helping butterflies has been tried for many years, but with limited success. Only recently have conservationists gained sufficient scientific knowledge to do this at all effectively and much research has still to be done on individual species.

Effort should be concentrated on re-establishing endangered species at well managed conservation sites where natural populations have died out in recent times - especially if there is a realistic potential for the species to remain permanently.

Conversely, attempts to establish species outside their accepted ranges are invariably unsuccessful and at best colonies die out after a few years.

- A leaflet setting out Butterfly Conservation's *Policy and Code on Butterfly Releases* is available on application.
- *A Code of Conservation Practice for Insect Re-establishment* is published by the Joint Committee for the Conservation of British Insects, copies of which are obtainable from the Royal Entomological Society of London.
- *Review of Butterfly Introductions* (Matthew Oates, 1990) published by World Wildlife Fund.

Code of practice for butterfly releases

- **All releases of butterflies, or of their earlier stages, must be planned properly and serve a clear conservation purpose. Releases should be both scientifically desirable and necessary. Butterflies should not be treated as amenity objects and releases should not be for publicity purposes.**
- **Liaise with local wildlife organisations. Avoid indiscriminate releases which can ruin on-going monitoring on valuable sites.**
- **The planned receiving site must be fully suitable for the species and maintainable as such. Detailed management plans are essential.**
- **Permission must be obtained from the owners of both the receiving site and the donor site. If either is a Site of Special Scientific Interest prior permission must be obtained from English Nature.**
- **Only native stock of indigenous species may be released into the wild. Releases of non-native species, sub-species or even foreign stock will contravene the Wildlife and Countryside Act 1981.**
- **The release of a viable number of individuals at the most appropriate life-stage must be ensured. The correct formula varies from species to species.**
- **Use of wild stock is preferable from a local source if possible, ideally, from the nearest sizeable colony. Particular care should be taken not to weaken the parent population, and to avoid the transfer of individuals during seasons of scarcity.**
- **All establishment attempts must be closely monitored and recorded. In some cases this will involve monitoring an immature stage. Much useful knowledge can be gained from clear information on such attempts even when unsuccessful.**
- **Establishments conducted in the name of Butterfly Conservation must conform with the Society's Policy on Butterfly Establishments.**

The Heath Fritillary relies on regular coppicing in woodland.

An Essex wood where the Heath Fritillary has been re-established from Kent stock. A plentiful supply of Cow-wheat, seen flowering in the foreground, is essential for the butterfly's survival.

LEARNING MORE ABOUT BUTTERFLIES

"Eat, drink and satisfy your desires as the butterfly, who partakes of the flower without stealing its fragrance or harming its tissue".

Watching butterflies in their natural surroundings is a delightful pastime. Moreover, it is the best way to understand where and when they may be seen as well as gaining a knowledge of their behaviour and life cycle. Searching for butterfly eggs and caterpillars will develop an insight into nature itself.

The London Butterfly House, Syon Park showing the 'outdoor' section for British butterflies.

OBSERVING WILD BUTTERFLIES

The beginner should explore the countryside, see butterflies or better still join a society such as Butterfly Conservation which organises field trips to the best butterfly spots. Led by experts, you will quickly learn how to look for butterflies and identify different species. The ability to recognise suitable butterfly habitats, larval foodplants and nectar flowers is an essential skill for the conservationist.

The butterfly 'season' depends largely on the weather and air temperatures. Hibernating butterflies may appear briefly on any warm day but the season usually starts in late April to mid-May onwards. The summer months of June, July and August are best for observing butterflies with all our species on the wing at some time during that period. Many will survive on through September and even into October.

Butterflies love the sun and will fly only during daylight hours. Most activity takes place between the hours of 10.30 am to 4.30 pm (BST). Even in cloudy weather an air temperature of 17°C is sufficient. Below this sunshine is necessary for flight and they are unlikely to be seen at 14°C or if it is windy or raining.

A Butterfly Conservation field trip to Ivinghoe Beacon, Bedfordshire. The site was cleared of hawthorn scrub and is now recolonised by the Duke of Burgundy.

BUTTERFLY BOOKS

Learning about butterflies is not just an outdoor summer activity as some easy reading will help prepare for fieldwork. For the identification of butterflies undoubtedly the best book is Jeremy Thomas's *Guide to British Butterflies*, published by Hamlyn. *Butterfly Watching*, by Paul Whalley, gives an easily readable and informative insight into making the most of watching butterflies in the wild.

There is a vast range of both old and modern literature about butterflies. The formation of a library of butterfly books is an informative and absorbing hobby.

BUTTERFLY HOUSES

Across the country some 40 or more butterfly 'houses' or 'farms' have sprung up over the last ten years. These commercially run ventures are open to the public during the summer months and a few throughout the year. They consist of large enclosed areas - usually horticultural greenhouses - with carefully controlled heating, lighting and humidity. There one can wander amongst exotic tropical plants and flowers with butterflies flying freely around.

Some foreign species will breed in these conditions but others are imported in pupa form from the Far East and Central or South America. A few butterfly houses have set up enclosed outside areas for British and European species.

A visit to a butterfly farm is an enjoyable trip for all the family, especially children. Butterfly Conservation keeps a list of Butterfly Houses which is available on application. Some are closed during winter. Always check opening times before going to a Butterfly House.

SCHOOLS AND BUTTERFLY CONSERVATION

From an early age children have a natural fascination for butterflies. A butterfly theme offers an easy and absorbing subject for study in the classroom and outdoor projects. There are many activities which can be incorporated into the school curriculum and are easily prepared by teachers. In addition to the educational value, pupils can learn an awareness and an understanding of wider conservation issues.

The Butterfly Conservation *Education Pack* is designed for the age range from First School to GCSE and A-level developing transferable skills. Contents include a mass of ideas and suggestions. Also useful is *Conservation in School Grounds* (BTCV 1988).

Butterfly Conservation's Education Pack for schools

- Butterfly biology and the life cycle
- Conservation projects
- Butterfly arts and crafts ideas
- Breeding butterflies
- Butterflies in poetry and prose
- School wild garden for butterflies
- School butterfly house
- Garden butterfly survey
- Butterfly information - books, societies, seeds, periodicals
- Butterfly field-study project

Available from Butterfly Conservation.

PHOTOGRAPHY

Many people have forsaken the butterfly net for the camera. The excitement and satisfaction of photographing beautiful butterflies at close quarters is an ideal substitute to the hobby of collecting them.

Photography is an easy way of documenting habitat changes both on a seasonal basis and, more importantly, over a longer period of years. It is surprising how quickly sites become overgrown or shaded out if they are neglected. Pictures should be taken regularly from the same position and dates carefully recorded.

The advent of advanced camera systems with auto-exposure and auto-focus, and in particular the single-lens reflex camera, has brought close-up butterfly photography well within the capabilities (and limited budget) of most butterfly enthusiasts. A variety of camera systems can be used with colour print or slide film to take dazzling pictures of butterflies and their early life stages.

From an early age children can enjoy the fascination and beauty of butterflies.

Best results for close-up work are with a macro lens of between 50mm and 100mm. This will allow close-focusing to obtain images up to half life-size or even more. A mono-pod helps to reduces camera shake (a major problem with close-ups). Dual flash is favoured by some photographers, although it can be rather cumbersome.

Pictures of the habitats and surrounding countryside where the butterflies were seen, will add to the value and interest.

Such quality results enable comprehensive visual records which a few years ago would have been inconceivable. Photography is undoubtedly an absorbing pastime where the best results will only be achieved through practice and patience, especially when working in the wild.

Video filming is relatively untried by butterfly enthusiasts. After the initial outlay on equipment it is a cheap way of recording butterflies, their development and habitats. Slide photographs can be transposed onto video without difficulty and the earlier work of a photographer taking 'stills' is not wasted by a change to video.

One word of warning. Photographers must be careful not to damage habitats by trampling undergrowth or breaking down trees and shrubs. Stay on paths and observe the country code.

• Butterfly Conservation is soon to publish a comprehensive booklet entitled *Butterfly Photography*.

PUTTING BUTTERFLIES ON THE MAP

An early summer transect walk. Butterfly recording does not require great skill. Learning to recognise similar species, such as the Small and Essex Skippers, needs practice and good eyesight.

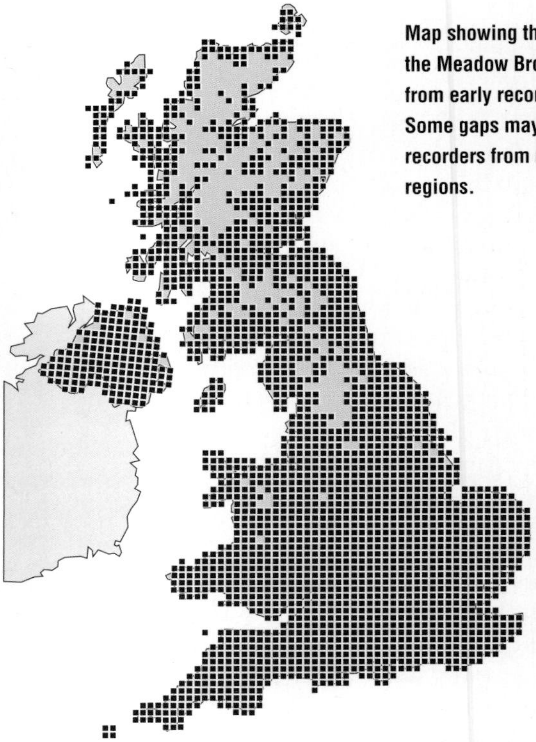

Map showing the distribution of the Meadow Brown compiled from early records up to 1988. Some gaps may reflect lack of recorders from more isolated regions.

The conservation of butterflies can only be effective if there is a clear picture of the distribution of each species and the status of populations from year to year.

SURVEYING AND MONITORING

Regular up-dating of the national distribution maps gives a striking image of the changes which occur both in the short and long-term. These maps and those compiled at county level, together with the monitoring of butterflies at individual sites, enable declines to be noted and steps taken to identify and reverse the causes. Occasional increases can also be pinpointed.

The continuous recording of butterflies across the country is a huge task and relies very largely on the enthusiastic work of amateur observers. Even those with only limited knowledge and experience can easily play a useful part in collecting records whether for the national scheme or a local project.

In recent years the national records have proved invaluable to scientists researching threatened species. And at local sites butterfly counts have highlighted the need for active habitat management and, when this has been implemented, the measure of success can be traced.

THE NATIONAL RECORDING SCHEME

This is organised by the Institute of Terrestrial Ecology at Monks Wood Experimental Station, Huntingdon, and was started at their Biological Records Centre (BRC) in 1967. The information collected was used in the publication of the *Atlas of Butterflies in Britain and Ireland* (1984) and more recently in the *Moths and Butterflies of Great Britain and Ireland* (1989, Harley Books). The BRC will supply instructions and recording cards to anyone wishing to participate in the recording scheme.

The ten kilometre square is the unit of recording, each dot on the map indicating that at least one record has been received from the area represented by the square. A limitation is that it is difficult to know if the dot is for a single sighting or a large colony of butterflies. However, Butterfly Conservation has devised an improved system whereby different shaped dot symbols are used to denote time periods. Even so this does not solve the problem of showing numbers. Contractions in range can be indicated but not expansions, as the most recent recording is given in preference.

Butterfly Conservation is currently seeking to co-ordinate the collation of records and promote and publicise the development of the National Recording Scheme.

The objective is to establish a computer data base with a dequate capacity and resources for input and retrieval. There is a need for a minimum but standard recording system for all recorders with emphasis on targeting rarer and endangered species at national level. The resulting 'Atlas' of Butterflies will be up-dated every 10-15 years, perhaps with limited or informal revisions published every 5 years.

Sensitive information relating to the location of

This butterfly site recording form (and the one on the following page) can be enlarged to A4 on a photocopier and used to contribute to the Butterfly Conservation recording schemes. Alternatively, copies can be obtained direct from Butterfly Conservation.

Butterfly site recording form

Year________

BUTTERFLY CONSERVATION

Dedicated to saving wild butterflies and their habitats

Name ______________________________

Telephone______________

Address ______________________________

____

____

Please use separate forms for different sites, or for different sub-sites within a large site. Please do not put data for more than one year on the same form.

Site information: if possible, please add a sketch map of the site and the area visited

Site name ______________________________ Grid ref ______________

Nearest town/ village______________________________ County ______________

Habitat types: Please tick the box(es) that apply to the area visited and give any other details about habitat and land use

- ☐ **2** Freshwater edges (lakes/ rivers/ canals)
- ☐ **31** Heath/ scrub
- ☐ **34** Dry calcareous grassland/ chalk down
- ☐ **38** Meadow (unimproved)
- ☐ **41** Broad-leaved deciduous woodland
- ☐ **43** Mixed woodland
- ☐ **54** Bog/calcareous fen
- ☐ **81** Fertilised/ improved/ reseeded grassland
- ☐ **82** Crops
- ☐ **83** Orchards/ plantations/ commercial forestry
- ☐ **84** Tree lines/ hedges
- ☐ **85** Parks/ gardens/ churchyards
- ☐ **86** Urban/ industrial areas
- ☐ **87** Fallow/ waste land
- ☐ **89** Quarries/ chalk pits/ gravel pits/ clay pits
- ☐ **90A** Road/ rail verges/ cuttings etc. (active)
- ☐ **90D** Road/ railway tracks (disused)

Other details

____

Comments

How to use this form

This sheet is for recording your details and information about the site habitat. The second sheet (or the reverse of this one if backed up) is for recording information about each visit.

When complete please send to: Butterfly Conservation, PO Box 222, Dedham, nr. Colchester, Essex CO7 6EY

Site information

Site name: Enter the name of the site, or a geographical feature as used on the OS map or a known local name. This could be a nature reserve, the name of a wood, or a well defined feature such as a green lane, or a hill name, for example. For a garden, it helps to give the postcode, and for a suburban location, the name of the district. If possible, attach a sketch map of the site and area visited.

Nearest town: Give the name of the nearest town or village to assist with identifying sites with a very local name or a very common name (such as Home Farm). It also helps in correcting grid reference errors (which anyone can make).

Grid ref: Enter the six figure OS Grid reference of the centre of the named site or sub site (eg. Oakley Wood (Bernwood) is SP615117) or at least a four figure reference to the 1Km square (eg. SP6111). If you have difficulty with OS Grid references, please attach a sketch or other indication of the location of the site to assist us.

County: As data may be sent to county museums and will be collated nationally, it is very useful to identify the county to which these records relate.

Habitat types: Tick the boxes in the list that best describe the habitat(s) present on the site, within the area visited. Please add any additional information that you may have on the habitat, not covered by the categories list.

Comments: Any additional notes on the site (eg. ownership, status or unusual sightings). Continue on a separate sheet if necessary.

Visit information

Enter details of each visit to the site using one column per visit. Give the **day** and **month** of the visit at the top, the approximate **length** of the visit (in minutes) and an idea of the **weather:**

- **P** for poor (cloudy/ cool/ windy)
- **M** for moderate
- **I** for ideal (warm/ sunny/ calm)

Species seen

To be likely to observe all the species that may occur on any site, it is best to make **at least 4 visits** through the season, i.e. April/May to September, but if you can only manage one visit please record the data anyway. For each species observed on a visit, enter the code for number seen as follows:

- **A** for one only
- **B** for 2 to 9
- **C** for 10 to 29
- **D** for 30 to 99
- **E** for 100 or more

If you see any eggs, larvae or pupae, please enter the codes **O**,**L**,or **P** respectively. If you see mating pairs enter the code **M**.

Butterfly site recording form

Details of site visits																						
Day																						
Month																						
Length of visit (mins)																						
Weather (P, M, I)																						
Small Skipper																						
Essex Skipper																						
Lulworth Skipper																						
Silver-spotted Skipper																						
Large Skipper																						
Chequered Skipper																						
Dingy Skipper																						
Grizzled Skipper																						
Swallowtail																						
Wood White																						
Clouded Yellow																						
Brimstone																						
Large White																						
Small White																						
Green-veined White																						
Orange Tip																						
Green Hairstreak																						
Brown Hairstreak																						
Purple hairstreak																						
White-letter Hairstreak																						
Black Hairstreak																						
Small Copper																						
Small Blue																						
Silver-studded Blue																						
Northern Brown Argus																						
Brown Argus																						
Common Blue																						
Chalkhill Blue																						
Adonis Blue																						
Holly Blue																						
Duke of Burgundy																						
White Admiral																						
Purple Emperor																						
Red Admiral																						
Painted Lady																						
Small Tortoiseshell																						
Large Tortoiseshell																						
Peacock																						
Comma																						
Small Pearl-bordered Frit.																						
Pearl-bordered Fritillary																						
High Brown Fritillary																						
Dark Green Fritillary																						
Silver-washed Fritillary																						
Marsh Fritillary																						
Glanville Fritillary																						
Heath Fritillary																						
Speckled Wood																						
Wall Brown																						
Scotch Argus																						
Mountain Ringlet																						
Marbled White																						
Grayling																						
Gatekeeper																						
Meadow Brown																						
Ringlet																						
Small Heath																						
Large Heath																						

endangered species would only be released to bona fide organisations or individuals. Recording by county or local trusts is crucial for particular conservation projects and records need to be uniform to dovetail with the national scheme.

A programme for the Co-ordination of Butterfly Recording in Britain and Ireland is a report published by a working party and the Biological Records Centre, 1992.

Local and county recording schemes

Most local Branches of Butterfly Conservation now assemble records for the area covered by the Branch. Very often the County Wildlife Trust or equivalent will be involved in a similar exercise and clearly there should be close liaison to share information and ensure the best use of resources especially to avoid duplication of effort. Large national landowners, such as the National Trust, invariably welcome the opportunity to work with local groups to survey their estates and co-operate on conservation issues.

Some county recording schemes have resulted in the publication of excellent books which are invaluable to conservationists. Most notable are undoubtedly *Butterflies of Dorset* (1984, Thomas and Webb), *Butterflies of Hertfordshire* (1987, Sawford) and *Butterflies of the London Area* (1987, Plant) amongst a number of others

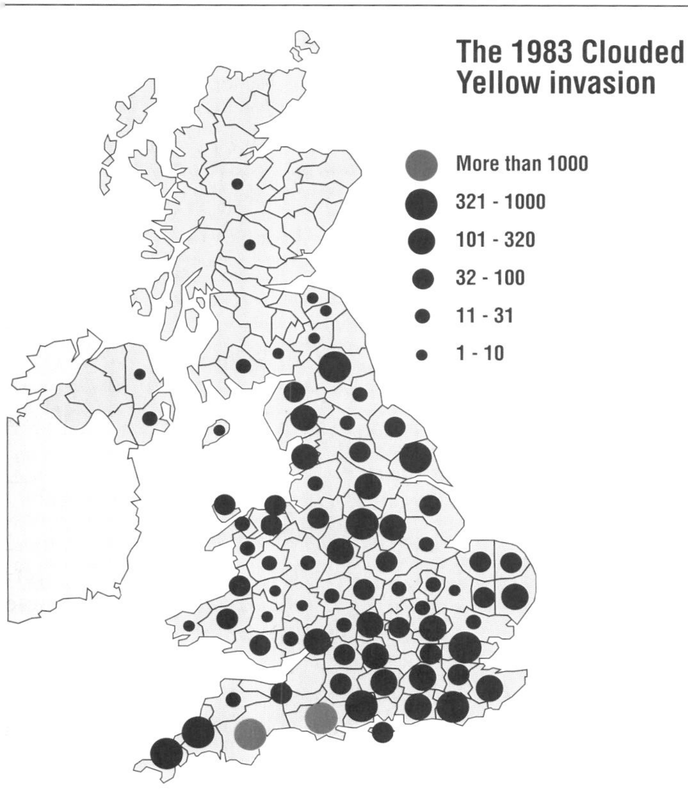

This map shows the numbers of Clouded Yellows recorded in each county during the 1983 'invasion' of migrants. Different shaped or coloured symbols can be used in a number of ways to present information.

A success story: the Comma was once restricted to a few counties on the Welsh borders, but has dramatically expanded its range throughout England and Wales. Such an expansion is easily represented on a map.

Larger scale maps are used and a much more detailed picture of distribution can be shown. A bibliography of current county lists (not necessarily with maps) is contained in *Butterfly Conservation News* No. 45. A full catalogue of all published material is found in *Local Lists of Lepidoptera* (1989, Chalmers-Hunt).

Records of all butterflies are important even if they are common species at well-known sites. One-off or regular visits are equally valuable and it should never be assumed that someone else has sent in a record. The chances are that this is not so. Sightings in the garden, your home town or village, or whilst on holiday all help to build up a picture and contribute to the local and national schemes.

MONITORING BUTTERFLY POPULATIONS

By studying butterfly distribution maps much can be learnt about their ecology and the factors which control their range when comparisons are made with maps of geology, climate and foodplant distribution. There is, however, a difficulty in representing on single maps the fluctuations in populations or actual number of butterflies and more detailed methods are needed to study these factors on local sites.

A standard method of regularly recording butterfly numbers is essential in planning and monitoring conservation projects on specific sites. Active management of habitats can be assessed in relation to the resultant increase or decrease in the numbers and location of butterflies. Butterflies respond very rapidly to changes in their habitat and are a good indicator of the success or otherwise of a management regime.

Two people visiting a site may form very different opinions about the number of butterflies they see, depending on the time spent looking, the weather, the area they cover, and their subjective comparisons with numbers of butterflies observed elsewhere. Standard methods of recording have, therefore, been devised and are briefly described in the following section.

The Countryside Code

- **Enjoy the countryside and respect its life and work**
- **Guard against all risk of fire**
- **Fasten all gates**
- **Keep your dogs under control**
- **Keep to public paths**
- **Use gates and stiles to cross fences, hedges and walls**
- **Leave livestock, crops and machinery alone**
- **Take your litter home**
- **Help keep water clean**
- **Protect wildlife, plants and trees**
- **Avoid trampling plants**
- **Take special care on country roads**

THE TRANSECT WALK

This involves the regular weekly walking of a fixed route between April and September. The walk

Garden butterfly survey form

Year________

Name ______________________ Telephone ______________________

Address ______________________

BUTTERFLY CONSERVATION

Dedicated to saving wild butterflies and their habitats

Please do not put recordings for more than one year on the same form, or combine your observations from different gardens.

Location of garden

Address (if different from above) ______________________

Grid ref ______________________

Postcode ______________________

Description of garden: Please tick the boxes that apply to your garden. Note any special features in the space below.

Is the garden environment:

- ☐ Urban
- ☐ Suburban
- ☐ Rural

Is your garden:

- ☐ Large
- ☐ Medium
- ☐ Small

Is your garden:

- ☐ Mainly open to the sun
- ☐ Mainly shaded
- ☐ Partly shaded, partly open

Does your garden face:

- ☐ North
- ☐ South
- ☐ East
- ☐ West

Is your garden:

- ☐ Mainly sheltered from the wind
- ☐ Mainly exposed to the wind
- ☐ Partly sheltered, partly exposed

Does your garden contain:

- ☐ Ornamental flowers and shrubs
- ☐ Wild flowers and grasses
- ☐ Mature fruit trees
- ☐ Vegetables

Proximity of wild habitats: Please tick any boxes that apply to your garden and give details.

Is your garden near to:	Approx. distance from your garden:	Terrain between wild habitat and garden:
☐ Woodland		
☐ Unimproved grassland		
☐ Heathland		
☐ Farmland		
☐ Wasteland		
☐ Other (please specify below)		

Comments

How to use this form

This sheet is for recording details of your garden.The second sheet (or the reverse of this one if backed up) is for recording information about butterfly observations during the year.

Recording details

Please record weekly sightings on part 2 of the form. Weeks run from Sunday to the following Saturday. If sightings are for only one day in a week, write in the date for that day. If relevant also give an indication of the weather conditions:

- **P** for poor (cloudy/ cool/ windy)
- **M** for moderate
- **I** for ideal (warm/ sunny/ calm)

Species seen

For each species observed, enter the code for numbers seen as follows:

- **A** for one only
- **B** for 2 to 9
- **C** for 10 to 29
- **D** for 30 to 99
- **E** for 100 or more

If you see any eggs, larvae or pupae, please enter the codes **O**,**L**,or **P** respectively. If you see mating pairs enter the code **M**.

When complete please send to: Butterfly Conservation, PO Box 222, Dedham, nr. Colchester, Essex CO7 6EY

is mapped and the number of butterflies seen within a certain distance is noted for collation at the end of each season so comparisons can be made from year to year.

Although the walk is not a direct count of butterfly populations a consistent recorder will provide a revealing commentary on the fluctuations in numbers of each species.

A transect can be set up anywhere if butterfly records are needed - nature reserves, fields, woods or a combination of habitats. A circular walk is often more convenient. A walk should not be too long or else it can become a chore. About twenty minutes or so is about enough and is a pleasant and rewarding way of spending a short while looking at butterflies each week.

Canopy dwellers or inconspicuous butterflies (especially the Hairstreaks) do not lend themselves well to the transect method and more suitable counting methods can be devised for them. This may involve counting eggs or larvae, which are easier to find than adults, or counting the number of flights of adults in a fixed time from a particular tree or section of hedgerow.

This butterfly site recording form (and the one on the following page) can be enlarged to A4 on a photocopier and used to contribute to the Butterfly Conservation recording schemes. Alternatively, copies can be obtained direct from Butterfly Conservation.

The important thing is to make careful notes of the exact location, time and method used and to repeat the counts under exactly the same conditions, otherwise comparisons will not be valid.

An exact record of methods used will also enable others to repeat the exercise at a later date.

A booklet entitled *Butterfly Monitoring Scheme - instructions to Independent Recorders* is available from the Institute of Terrestrial Ecology.

Butterfly Conservation garden habitat survey

A very simple system is used to record species observed in a garden or any other single habitat, even in towns. The records are being accumulated with some interesting and unexpected results. Details from Butterfly Conservation.

Counting methods can be adapted to suit particular species and a combination of the different systems devised. Larva and egg counts may also assist in estimating numbers and annual changes. The marking of butterflies should only be undertaken by experts and is of limited value unless it is part of a professional research project.

Butterfly recording is a crucial part of conservation as well as being an enjoyable and rewarding pastime. Everyone can contribute to recording schemes which will only be successful with the help of enthusiasts across the country.

Garden butterfly survey form

Month
Week beginning
Single day observation
Weather (P, M, I)
Small Skipper
Essex Skipper
Lulworth Skipper
Silver-spotted Skipper
Large Skipper
Chequered Skipper
Dingy Skipper
Grizzled Skipper
Swallowtail
Wood White
Clouded Yellow
Brimstone
Large White
Small White
Green-veined White
Orange Tip
Green Hairstreak
Brown Hairstreak
Purple hairstreak
White-letter Hairstreak
Black Hairstreak
Small Copper
Small Blue
Silver-studded Blue
Northern Brown Argus
Brown Argus
Common Blue
Chalkhill Blue
Adonis Blue
Holly Blue
Duke of Burgundy
White Admiral
Purple Emperor
Red Admiral
Painted Lady
Small Tortoiseshell
Large Tortoiseshell
Peacock
Comma
Small Pearl-bordered Frit.
Pear-bordered Fritillary
High Brown Fritillary
Dark Green Fritillary
Silver-washed Fritillary
Marsh Fritillary
Glanville Fritillary
Heath Fritillary
Speckled Wood
Wall Brown
Scotch Argus
Mountain Ringlet
Marbled White
Grayling
Gatekeeper
Meadow Brown
Ringlet
Small Heath
Large Heath

Section Four BUTTERFLIES UNDER THREAT

"We seem to have lost that ancient kinship with nature that not so long ago was instinctive to us"

Britain has 25 species of butterfly considered to be in danger of extinction, facing serious reduction in numbers and distribution or vulnerable for other reasons. All receive at least some legal protection under the Wildlife and Countryside Act 1981. However, this relates only to the butterflies. Their habitats do not necessarily enjoy any protection or a safe future.

Habitat protection, expansion and even creation are of crucial importance if these butterflies are to survive and flourish in years to come. Here we outline the current distribution of these butterflies and illustrate the losses recorded during the last two centuries. Habitats are described briefly along with the conservation requirements to keep strong colonies thriving.

Whilst much of the information contained in the habitat sections of this book will be helpful, they are unlikely to cover everything relevant to particular species of rare or endangered butterflies. As will be seen most of these have highly specific habitat requirements. Conservation work, therefore, involves specialist knowledge and careful planning which should be undertaken with the help of butterfly experts.

This section includes commentaries on all threatened species – except for the Large Blue and Large Copper, which have been re-introduced using foreign livestock. The following terms are used to describe the status of butterfly species:

- *ENDANGERED means it is in danger of extinction due to very low numbers and/or the small number of surviving colonies.*
- *VULNERABLE means the species has very precise habitat requirements which, if lost, will result in the disappearance of the butterfly from that location.*
- *RARE means very few around, uncommon at the best of times.*
- *WIDESPREAD means commonly found across a defined geographical area, either generally, or in particular habitats.*

BUTTERFLIES MOST AT RISK

Chequered Skipper ▲

Carterocephalus palaemon

Distribution: Extinct in England and now only found in western Scotland living in small, sometimes isolated colonies.
Habitat: Sheltered sunny lowland areas of rough scrub with grasses and plentiful nectar flowers.
Conservation: About 40 known colonies. Vulnerable but no immediate danger. Recent extinction in England necessitates monitoring of sites and numbers. Commercial afforestation threatens to destroy habitats. Tree growth must be controlled to maintain open sites or suitable rides and clearings. Re-introduction in England under consideration, but not many sites in suitable condition.

Lulworth Skipper ▼

Thymelicus acteon

Distribution: Concentrated on southern coastal areas of Dorset especially around Lulworth with scattered colonies along Devon and Cornish coast.
Habitats: Warm south-facing chalk or limestone hillsides and cliffs with longer grasses.
Conservation: Its very specific habitat requirements appear to have benefited from reduction of rabbit grazing due to myxomatosis and domestic grazing. Long uncut or ungrazed grasses favoured for egg laying and larval development. Some colonies have expanded over the last decade; nevertheless it is vulnerable to changes in its grassland habitats.

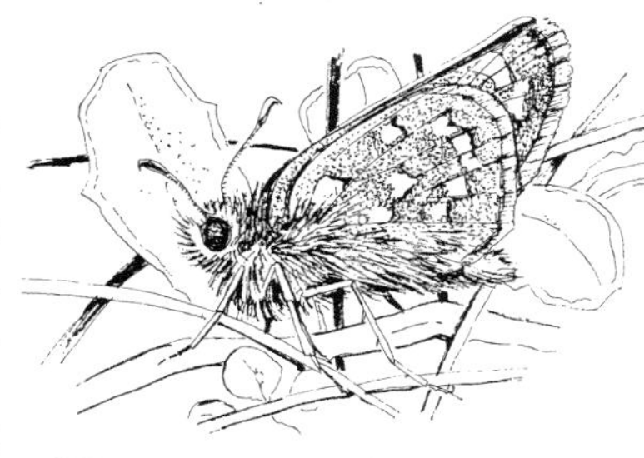

Silver-spotted Skipper ▲

Hesperia comma

Distribution: Formerly more widespread but never common, it is now restricted to only about 50 isolated colonies in southern England.
Habitats: Steeper south facing chalk hillsides with thin topsoil and bare earth or chalk fragments.
Conservation: Maintenance of short grasses and vegetation on suitable sites by occasional light grazing. Size of colonies limited by extent of suitable tufts of sheep's fescue grass on which eggs are laid. Small patches of bare ground are essential. Very vulnerable to further decline.

Swallowtail ▶

Papilio machaon

Distribution: Norfolk Broads only, formerly more widespread in fenlands of Cambridgeshire and Lincolnshire. Eighteenth and 19th century writers suggest it was found across southern England but probably this was the European sub-species which has become extinct except for occasional migrants. The British Swallowtail is unique in that it has its own foodplant and differs in appearance to the continental variety.
Habitat: Restricted to the fens where foodplant milk-parsley grows in suitable, wet conditions. Many of the best locations for the Swallowtail are specially managed reserves or SSSIs.
Conservation: Fully protected under Wildlife and Countryside Act 1981. Maintenance of the water table at an adequate level to sustain sufficient quantity and quality of Milk-parsley is essential. Traditional cutting and harvesting of sedges and reeds prevents it being choked by coarse vegetation and scrubbing

over. Excess agricultural nitrates and phosphates in water may cause deterioration of vegetation. Current population has been quite strong in recent years and without any apparent threat provided the habitat does not alter.

Wood White ▲

Leptidea sinapis

Distribution: Scattered local colonies across the southern half of central and western England and Wales; more frequent across Ireland. Range contracted during last century but has marginally expanded again to the present day.
Habitat: Most favourable habitats are woodland clearings, rides and scrubby areas. Prefers slightly shaded conditions where cleared or coppiced woodland is beginning to regenerate. Tends to fly in more open country in Ireland.
Conservation: Rides or clearings should be cut or widened to prevent over-shading but give not too much sunlight. Coppicing cycle is likely to be very beneficial. Numbers fluctuate considerably due to annual climatic and temperature variations. Colonies may be threatened or reduced to dangerously low numbers by collectors.

Brown Hairstreak ▼

Thecla betulae

Distribution: Has declined steadily during last hundred years, and now concentrated in Wealden districts of Surrey and West Sussex, north Devon and Somerset borders, south west Wales and around the Burren in west Ireland. Also a few scattered colonies across central southern England. Extinct in Scotland and northern England.
Habitat: Wood-edges and tall high-density hedgerows enclosing small fields or more open but hilly countryside.
Conservation: Protection and maintenance of preferred areas especially sheltered low-lying wood-edges and hedges with widespread abundance of blackthorn foodplant. Hedge trimming during winter can destroy a large number of eggs. Further loss of isolated colonies likely but stronghold districts not thought to be in immediate danger.

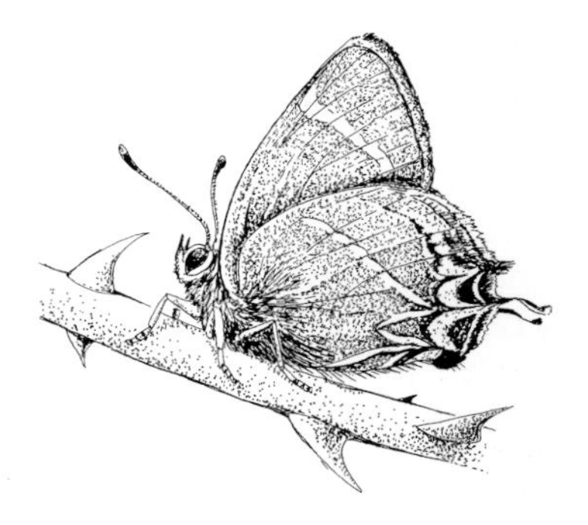

White-letter Hairstreak ▲

Satyrium (Strymonidia) w-album

Distribution: Found in scattered localities across most of England and Wales to a line from Lancashire to Durham. Less common in the West Country and central Wales.
Habitat: Small colonies now rely on elm sucker growth along wood-edges and rides of deciduous woods although the butterfly may fly in more open areas.
Conservation: As a result of Dutch elm disease the butterfly undoubtedly suffered a reduction in numbers. Extensive surveys have, however, revealed that this butterfly is probably more widespread than originally thought and many more colonies have been discovered. In fact it does not seem to be seriously threatened, the larva having the adaptability to feed on young elm shoots growing from old stumps which must be protected for the suckers. The re-establishment of elms, especially wych elm, and the preservation of woodland will assist the species to survive.

Black Hairstreak ▼

Satyrium (Strymonidia) pruni

Distribution: Essentially restricted to a small band across the east Midlands between Oxfordshire and Peterborough. Upwards of 30 sites still support colonies which is about half the number historically recorded. A few other colonies are scattered across southern England but these are suspected of being 'introduced' colonies and their long-term viability thought to be suspect.
Habitat: This butterfly has critical habitat requirements. Colonies are discrete and sedentary, choosing wood-edges and hedges with good concentrations of blackthorn or other Prunus. Although colonies tend to occupy relatively small areas averaging about 200 square metres they are usually located in woods of several hundred acres. Sunny sheltered locations tend to support larger numbers and are preferred breeding sites.
Conservation: A mixture of older blackthorn (20-60 years) and replacement coppiced

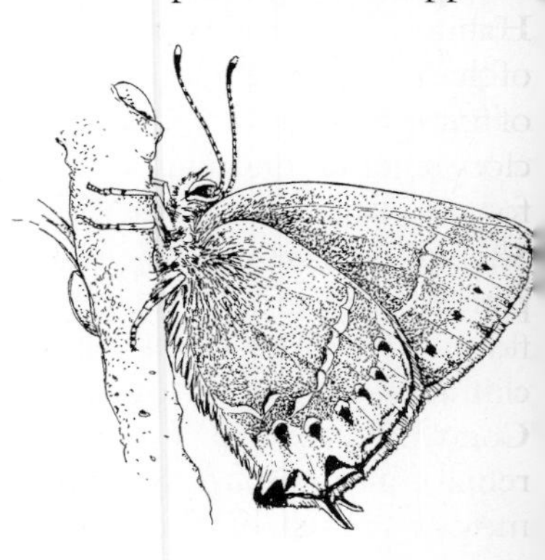

shrubs with a southerly aspect need to be maintained in suitable woodland or semi-open areas. Clearance of long grass and undergrowth at the base of blackthorn reduces predation of larva by mammals or game birds.

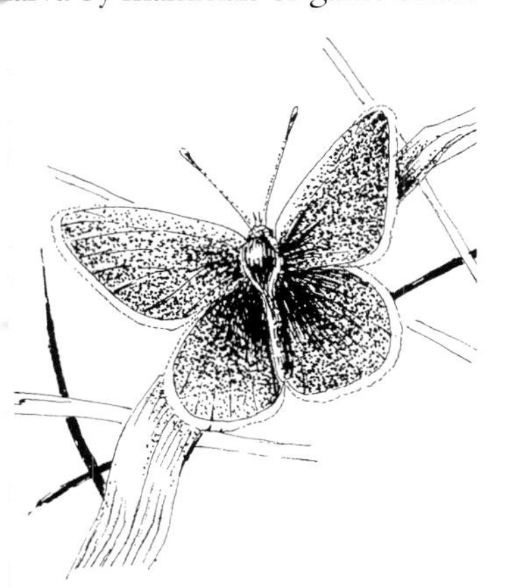

Small Blue ▲

Cupido minimus

Distribution: Very local isolated colonies exist mostly in the coastal regions of northern England, Wales, Scotland and Ireland. Elsewhere its distribution is concentrated in central and southern England. Virtually absent from the West Country and East Anglia.

Habitat: The main requirement of this butterfly is a good growth of flowering kidney vetch located close to long grass and shrubs for roosting and sun-basking. Sheltered sites seem to be preferred but sometimes colonies flourish on exposed hillsides or clifftops.

Conservation: Colonies tend to remain static and isolated, moving only when the kidney vetch has insufficient flowers. The foodplant must be maintained on the site or near enough for the butterfly to reach. Natural plant succession and over-grazing are the greatest threats but these can be controlled by simple management.

Silver-studded Blue ▼

Plebejus argus

Distribution: Sadly this butterfly has become extinct in northern England and Scotland over the last hundred years. Scattered colonies continue to survive across central England and Wales. More frequently found in heathland areas of central southern England and the West Country.

Habitat: Warm or south-facing lowland heaths with fresh heather growth, also surviving on one locality of limestone grassland.

Conservation: Natural deterioration or destruction of habitats has decimated this species. This has to be halted on remaining habitats. Heathland must be protected and maintained by traditional methods of burning or clearance. Agricultural management practices must be carefully employed on grassland to prevent dominance of coarse grasses and bracken.

Northern Brown Argus ▲

Aricia artaxerxes

Distribution: Located in seven separate confined areas in northern England and more widely in the south and east of Scotland. This is a separate sub-species to the Brown Argus found commonly in southern Britain.

Habitat: Preference is for warmer south facing slopes on limestone outcrops. Colonies need a plentiful supply of common rock-rose, thyme and bird's foot trefoil, commonly found growing on well-drained, rich soils.

Conservation: Breeding sites require regular monitoring to ensure the optimum level of grazing or management control for a sufficient quantity of the larval foodplants to flourish.

Chalkhill Blue ▼

Lysandra coridon

Distribution: Confined to chalk and limestone grassland south of a line from the Wash to the Bristol Channel.

Habitat: The butterfly prefers sunny, south facing slopes of chalk and limestone downland. Colony size will depend on the supply of the larval foodplant Horse-shoe vetch, with the butterflies rarely straying far.

Conservation: Loss of habitat has undoubtedly reduced the number and size of colonies over the last fifty years and to an extent the distribution as well. Changes in land use (especially ploughing and heavy cattle grazing) have to be avoided. Grassland needs light grazing - sheep are best or rotational cutting - so that longer coarse grasses do not choke the horse-shoe vetch. The gradual return of rabbits after myxomatosis may contribute to the improvement of remaining or potential habitats provided they are not over-grazed.

Adonis Blue ▼
Lysandra bellargus

Distribution: Follows exactly the chalk downlands of southern England but only in very local colonies. These fell to about 75 in the 1970s but increased again to 150 in recent years.
Habitat: Open but warm south facing slopes with very short cropped grass and bare earth patches for sun-basking. The female butterfly will only lay eggs in such areas, due to optimum ground temperature requirements and possible relationship between the caterpillars and certain species of ants.
Conservation: Ploughing for cereals or insufficient grazing have resulted in loss or reduction of many habitats. Careful management is necessary to keep grass short and prevent incursion by taller plants or grasses. This is best achieved by reasonably intensive grazing of cattle or horses. On steeper slopes sheep may be more suitable in order to avoid soil erosion. Plentiful growth of larval foodplant, horse-shoe vetch, and nectar plants for the butterfly, is essential.

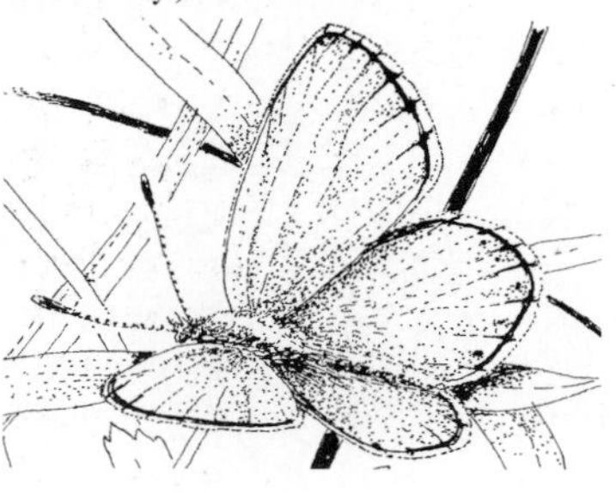

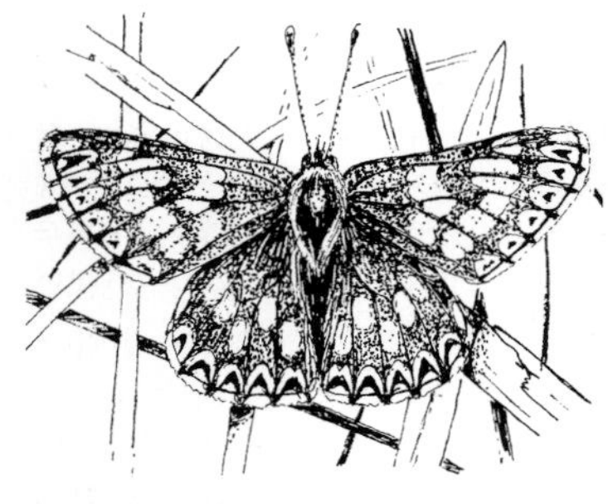

Duke of Burgundy ▲
Hamearis lucina

Distribution: Essentially southern central England with a few isolated populations in Yorkshire, Lancashire and Kent. Much more widely distributed in the last century, since when it has declined steadily both in numbers and distribution.
Habitat: Small isolated colonies occur mainly on sheltered downland slopes with tall scrub and longish grass or more rarely on regularly coppiced woodland.
Conservation: Woodland colonies are relatively scarce nowadays as a result of changes in management practices. The butterfly is sedentary so will only survive with active management enabling it to move just short distances from one site to another where the larval foodplant, primrose, is abundant. On downland habitats the foodplant is cowslip which may be excluded if coarser grasses or scrub are allowed to dominate. Light grazing in autumn or winter should be adopted to maintain medium height of grass and restrict scrub, especially hawthorn.

Purple Emperor ▼
Apatura iris

Distribution: Largely restricted to the central counties of southern England but still found in scattered locations around the adjoining counties, in East Anglia, the south Midlands, the Nottingham area, Somerset and Kent.
Habitat: Larger deciduous woodland with a preference for those with a predominance of old oak where they congregate around the tops of tall 'master' trees. May also be found around fringes and more open areas when dispersing.
Conservation: Has suffered from the felling of forests and planting of coniferous plantations for commercial growing. Hence the range of its distribution and numbers have reduced substantially over the last 100 years. Further loss or fragmentation of suitable woodland must be resisted. A continuous sequence of maturing sallows, the larval foodplant, should be encouraged along woodland rides and in glades. Where timber is cut, replanting should be with native deciduous trees, ideally oak.

Large Tortoiseshell ▲
Nymphalis polychloros

Distribution: During the last century this butterfly was apparently widely found in England with a few records from Wales and Scotland as well. Since then its appearance has been very erratic and apart from one period in the 1940s of relative abundance, may now be regarded as extinct in Britain. The pattern of recent sightings in southern counties indicates that the Large Tortoiseshell is an occasional migrant. As a highly mobile species, identifying locations where it might be seen has always proved unreliable.
Habitat: Woodlands and hedgerows but also gardens and elsewhere when the butterfly seeks nectar flowers for feeding.
Conservation: The reasons for this butterfly's decline are not at all certain. Dutch elm disease undoubtedly weakened the butterfly in its last strongholds but its decline started a long while before. Climatic changes, especially in humidity levels, and an increased susceptibility to parasites, may have contributed to its disappearance.

Pearl-bordered Fritillary ▼
Boloria euphrosyne

Distribution: Found on suitable sites across southern England and a fair number of localities throughout Wales; also in scattered areas across the rest of England and in Scotland and one confined part of western Ireland.
Habitat: Recently cleared woodland, especially large plantations, and to a lesser extent ride edges will support this butterfly. Further north it seems to prefer drier, more open habitats with light woodland or scrub.
Conservation: Many colonies have died out due to forestry management changes and cessation of coppicing. An abundance of dog violet, the caterpillar foodplant, or other violets, growing even in small areas of woodland, will support colonies. A succession of coppiced or cleared woodland is essential. The butterfly will move from one breeding site to another provided it can follow lines of sunny rides or paths.

High Brown Fritillary ▲
Argynnis adippe

Distribution: From early records this butterfly was found across England and Wales although colonies were limited by availability of suitable habitat areas. Since the 1950s it has declined rapidly to the extent that it is virtually extinct in all but a few counties in southwest England. Northwest England and Wales also have a few scattered colonies. Changes in woodland management practices, allowing succession to a closed tree canopy, are almost certainly the main cause of the loss. Changes in climatic conditions and poor weather during important development periods in spring may be contributory factors.
Habitat: Typical habitat is open woodland rides or clearings where uninterrupted sunshine reaches the ground all day. An abundance of violets must be present. Further west and north, colonies favour more open, scrubby hillsides.
Conservation: Precise requirements are still uncertain but the butterfly seems to prefer small scale areas of coppice mosaic, scrubby wood or glade edges or hillside with scrub or bracken. A sufficiently low undergrowth must be maintained for violets to grow plentifully. Direct sunlight and adequate ground temperature are crucial for the caterpillar to reach maturity. Incursion of tree canopy or bracken shading all the ground must be avoided.

Dark Green Fritillary ▲
Argynnis aglaja

Distribution: Whilst this butterfly is found throughout Britain it is concentrated mostly in the south and west, whereas elsewhere it prefers coastal areas or may only occur in isolated colonies. In Ireland it is fairly common right the way round the coast with only a few inland populations. The last few years have seen a dramatic but unexplained fall in numbers and the loss of some colonies altogether.
Habitat: Open calcareous downs, hills, cliffs and rough moorland appear to be the most favoured habitats, especially where grasses are long and dense. The butterfly can also survive in open woodland or scrub clearings but will die out if undergrowth or tree shade increases.
Conservation: The very recent decline of this butterfly is not understood but is likely to be a combination of climate changes and unsuitable land management. The caterpillars will feed on a variety of violet species. These foodplants must grow in sufficient abundance but, more importantly, be positioned acceptably in terms of humidity and temperature for the female to lay. Occasionally light grazing will help maintain suitable conditions but on no account should land be fertilised or ploughed.

Marsh Fritillary ▲
Eurodryas aurinia

Distribution: Many colonies formerly scattered across the British Isles but excluding high mountainous areas. Greatly reduced number of locations throughout the country during this century, and largely eliminated from eastern and midland areas now occurring mainly in the west, which is its stronghold.
Habitat: The Marsh Fritillary lives in very sedentary and

isolated colonies which may vary enormously in numbers. Sites must remain unshaded so the caterpillar can mature quickly. The most common habitats are damp fields, rough pastures, moorland or heathland. It also occurs on chalk downland especially in Wiltshire and Dorset.
Conservation: As the butterfly will not travel beyond its natural habitats, colonies are sedentary and cannot re-establish if they die out. Optimum ground cover must be maintained by light grazing. It should include a plentiful supply of Devil's-bit scabious for the caterpillars and nectar flowers for the butterfly.

Glanville Fritillary ▼
Melitaea cinxia

Distribution: Up to the mid-19th century this butterfly was described as widespread across southeast England. Since then it has declined and is now native only on the south coast of the Isle of Wight.
Habitat: The undercliffs along the coast where there are frequent landslips leaving bare earth for the foodplant ribwort plantain to regenerate.
Conservation: The reasons for the butterfly's decline are not certain. Climatic variations or land use changes may be contributory factors. Careful monitoring of remaining colonies and protection of habitats must be continued. Sea defences, cliff stabilisation, marinas and coastal development will destroy habitats. Collectors also pose a threat and must be controlled.

Heath Fritillary ▲
Mellicta athalia

Distribution: Once widespread but localised across southern England; but now very restricted and only found in one area of Kent and a few locations in the West Country. Due to the cessation of regular coppicing in woodland, the number of suitable sites has diminished and the butterfly has been on the verge of extinction. Concerted efforts by conservationists have helped prevent its final demise during the last twenty years.
Habitat: These vary but appear to fall into three distinct categories; grassland with plentiful Plantain, woodland clearings with abundance of cow-wheat and sheltered heathland with the same plant.
Conservation: Traditional coppicing of deciduous woodland will create a succession of suitable habitats. This has been successfully achieved in a Kent National Nature Reserve. On all habitat sites undergrowth and tree rejuvenation has to be checked in order to allow enough direct sunshine to reach the ground. Exact conservation management is still being researched and a re-establishment programme on an Essex reserve has shown promising results for a few years.

Mountain Ringlet ▼
Erebia epiphron

Distribution: Confined to the Lake District and the western Grampians in Scotland.
Habitat: Lives in small isolated colonies on grassy plateaux or in damp boggy hollows. In England the butterfly is only found between 500-700 metres altitude and in Scotland 350-800m.
Conservation: As a mountain butterfly, this species probably established itself during the retreat of the last Ice Age and has survived in isolated colonies for several thousand years. Little study exists of the conservation needs but, clearly, known sites must be allowed to remain undisturbed except for light grazing by sheep. Afforestation, cultivation or over grazing might be a threat in some instances.

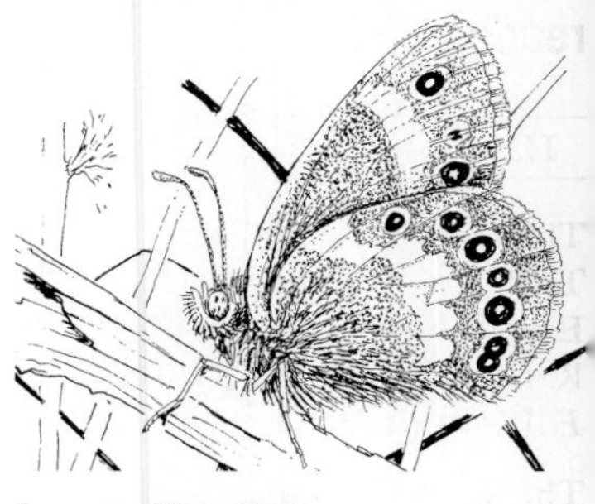

Large Heath ▲
Coenonympha tullia

Distribution: The Large Heath is a northern butterfly found commonly in Scotland but less so across Ireland. Scattered colonies survive in northern England and parts of north Wales. It has died out from many of its English sites due largely to land drainage, peat cutting and afforestation.
Habitat: Breeds in discrete colonies on flat, damp or boggy ground from sea level up to 800 metres. An abundance of mature cottongrass is needed, along with a suitable nectar plant such as heather or cranberry.
Conservation: Known sites supporting colonies need protection from drainage, peat cutting or other activity liable to reduce the water level. This urgently applies to surviving colonies in England and Wales if they are not to be destroyed altogether.

FURTHER INFORMATION

Bibliography and reading list

HISTORY AND GUIDES

The British Butterflies: Their Origin and Establishment
R. L. H. Dennis, 1977
E. W.Classey Ltd.

The Scientific Names of the British Lepidoptera: their History and Meaning
A. Maitland Emmet, 1991
Harley Books

Local Lists of Lepidoptera: a Bibliographical Catalogue of Local Lists and Regional Accounts
J. M. Chalmers Hunt, 1989
Hedera Press

Guide to the Butterflies of the British Isles
J. A. Thomas, 1986
Hamlyn

The Mitchell Beazley Pocket Guide to Butterflies *(European)*
P. Whalley, revised 1990
Mitchell Beazley

Butterfly Watching
P. Whalley, 1980
Severn House Publishers Ltd.

Butterflies of Britain & Ireland
J. Thomas &
R. Lewington, 1991
Dorling Kindersley

Butterflies and Day-Flying Moths of Britain and Europe
M. Chinery, 1989
Collins New Generation Guides

The Butterflies of Great Britain (Vol. 7, part 1 of the series)
A. Maitland Emmett,
J. Heath, 1990 revised
Harley Books

The Moths and Butterflies of Great Britain and Ireland (Vol. 7, part 2)
A. Maitland Emmet,
J. Heath, 1991
Harley Books

British Moths: A Complete Guide
M. Brooks, 1991
Jonathan Cape

Colour Identification Guide to Moths of the British Isles
B. Skinner, reprinted 1988
Viking, Penguin Group

British Pyralid Moths: A Guide to their Identification
B. Goater, 1986
Harley Books

The Ecology of Butterflies in Britain
R. L. H. Dennis, 1992
Oxford Science Publications

Butterfly Conservation
T. R. New, 1992
Oxford Science Publication

CONSERVATION

Butterflies of Europe, Aspects of the Conservation of Butterflies in Europe (Vol. 8 in a series)
Otakar Kudrna (Ed.), 1986
Aula-Verlag, Wiesbaden

Habitat Conservation for Insects: A Neglected Green Issue
R. Fry & D. Lonsdale, 1991
The Amateur Entomologists' Society

The Conservation of Butterflies Booklet
(NCC) 1981
English Nature

Land Management for Butterflies
1993
Butterfly Conservation

Legislation to Conserve Insects in Europe
N. M. Collins, 1987
The Amateur Entomologists' Society (Pamphlet no. 13)

Pesticides: Code of Practice for Safe Use on Forms and Holdings
Ministry of Agriculture, Food and Fishing, 1990
HMSO

A Guide to Countryside Conservation: Britain's Rural Heritage
Dr. John Feltwell, 1989
Ward Lock Ltd.

Butterfly Houses in Britain: The Conservation Implications
N. M. Collins, 1987
International Union for Conservation of Nature and Natural Resources

The Conservation of the Chequered Skipper in Britain
R. Collier, 1986
(NCC) English Nature

HABITAT MANAGEMENT

Habitat Management for Invertebrates
Peter Kirby, 1992
Royal Society for Protection of Birds

Management Plans: A Guide to their Preparation and Use
1986
Countryside Commission

The Management of Chalk Grassland for Butterflies
Butterflies Under Threat Team, 1986
English Nature

The Establishment and Management of Wild Flower Meadows
T. C. E. Wells, R. Cox
& A. Frost, 1989
English Nature

Wild Flower Grasslands from Crop-Grown Seed and Hay Bales
T. C. E. Wells, A. Frost
& S. Bell 1989
English Nature

The Conservation of Cornfield Flowers
1989 (booklet)
(NCC) English Nature

The Conservation of Meadows and Pastures
1988 (booklet)
(NCC) English Nature

The Management of Grassland and Heath in Country Parks
J. E. Lowday &
T. C. E. Wells, 1977
Countryside Commission

Hedging: A Practical Handbook
A. Brooks, Revised 1988
British Trust for Conservation Volunteers

Woodlands: A Practical Conservation Handbook
A. Brooks, Revised 1988
British Trust for Conservation Volunteers

Conservation in School Grounds
N. Forster, 1990
A 'Pack' published by BTCV

GARDENING

Gardening for Butterflies
Madge Payne, 1987
Butterfly Conservation

Garden Plants for Butterflies
Matthew Oates, 1985
Brian Masterson & Associates Ltd.

BREEDING

Breeding the British Butterflies
P. Cribb, 1983
The Amateur Entomologists' Society

Breeding the British and European Hawk-Moths
P. Sokoloff, 1984
The Amateur Entomologists' Society

Breeding Butterflies and Moths: A Practical Handbook for British and European Species
E. Friedrick, 1986
Harley Books

Keeping and Breeding Butterflies and Other Exotica
J. L. Stone, 1992
Blandford

RECORDING & SURVEYING

Monitoring Butterflies for Ecology and Conservation
E. Pollard & T.J. Yates, 1993
Chapman & Hall

Recent Surveys and Research on Butterflies: a Species Index and Bibliography
P. T. Harding & S. V. Green, 1991
Institute of Terrestrial Ecology

A Review of Butterfly Introductions in Britain and Ireland
M. R. Oates & M. S. Warren, 1990
WWF UK Ltd.

Monitoring the Abundance of Butterflies 1976-1985
(NCC) English Nature

INFORMATION & ORGANISATION

A Directory for Entomologists
D. Reavey & M. Colvin, 1989
The Amateur Entomologists' Society

Entomology: A Guide to Information Sources
2nd edition, P. Gilbert & C. J. Hamilton, 1990
Mansell

Organising a Local Conservation Group
BTCV 1988
British Trust for Conservation Volunteers

Useful addresses

SOCIETIES

Butterfly Conservation (British Butterfly Conservation Society Ltd)
P.O. Box 222
Dedham
Colchester
Essex CO7 6EY
Tel. 0206 322342

Amateur Entomologists' Society ***(Publications)***
The Hawthorns
Frating Road
Great Bromley
Colchester
Essex, CO7 7JN

The Butterfly and Moth Stamp Society
c/o 29 The Rising
Eastbourne
East Sussex BN23 7TL

Royal Entomological Society of London
41 Queen's Gate
London SW7 5HR
Tel. 071 584 8361

ENVIRONMENTAL ORGANISATIONS

Countryside Commission
Publications Department
19/23 Albert Road
Manchester M19 2EQ
Tel. 061 224 6287

Countryside Council for Wales
Plas Penrhos
Ffordd Penrhos
Bangor, Gwynedd
LL57 2LQ
Tel. 0248 370444

Scottish Natural Heritage
12 Hope Terrace
Edinburgh, EH9 2AS
Tel. 031 447 4784

British Trust for Conservation Volunteers
Publications
Conservation Practice Tools and Trading Ltd.
Bessemer House
59 Carlisle St East
Sheffield S4 7QN
Tel. 0742 755087

Forest Enterprise (formerly the Forestry Commission)
Publications Section
Alice Holt Lodge
Wrecclesham
Farnham
Surrey GU10 4LH

Institute of Terrestrial Ecology (ITE)
Monks Wood
Experimental Station
Abbots Ripton
Huntingdon
PE17 2LS
Tel. 04873 381
Fax: 04873 467

Farming and Wildlife Advisory Group (FWAG)
National Agricultural Centre
Stoneleigh, Kenilworth
Warwicks CV8 2RX
Tel. 0203 696699

English Nature (formerly part of Nature Conservancy Council)
Northminster House
Peterborough
PE1 1UA
Tel. 0733 897629

MAFF ***(Publications)***
Ministry of Agriculture, Fisheries and Food
London SE99 7TP
Tel. 081 694 8867

BOOKS & PUBLICATIONS

David Dunbar
31 Llanvanor Road
London NW2 2AR
Tel. 081 455 9612
Out-of-print, antiquarian and new reference books.

E. W. Classey Ltd
P.O. Box 93
Faringdon
Oxon SN7 7DR
Tel. 036782 399
New and secondhand entomological books.

Harley Books
Great Horkesley
Colchester
Essex CO6 4AH
Entomological book publishers.

OTHERS

Game Conservancy Trust
Fordingbridge
Hants SP6 1EF
Conservation research

The Natural History Museum
Cromwell Road
London SW7 5BD
Tel. 071 938 9123
Publications and museum specimen collections.

London Butterfly House
Syon Park
Brentford
Middlesex
Tel. 081 560 7272

CHECKLISTS OF ENGLISH AND SCIENTIFIC NAMES

Butterflies

Adonis Blue
Lysandra bellargus
Bath White
Pontia daplidice
Black Hairstreak
Satyrium (Strymonidia) pruni
Black-veined White
Aporia crataegi
Brimstone
Gonepteryx rhamni
Brown Argus
Aricia agestis
Brown Hairstreak
Thecla betulae
Camberwell Beauty
Nymphalis antiopa
Chalkhill Blue
Lysandra coridon
Chequered Skipper
Carterocephalus palaemon
Clouded Yellow
Colias croceus
Comma
Polygonia c-album
Common Blue
Polyommatus icarus
Dark Green Fritillary
Argynnis aglaja
Dingy Skipper
Erynnis tages
Duke of Burgundy
Hamearis lucina
Essex Skipper
Thymelicus lineola
Glanville Fritillary
Melitaea cinxia
Grayling
Hipparchia semele
Gatekeeper
Pyronia tithonus
Green Hairstreak
Callophrys rubi
Green-veined White
Pieris napi
Grizzled Skipper
Pyrgus malvae
Heath Fritillary
Mellicta athalia
High Brown Fritillary
Argynnis adippe
Holly Blue
Celastrina argiolus
Large Blue
Maculinea arion
Large Copper
Lycaena dispar
Large Heath
Coenonympha tullia
Large Skipper
Ochlodes venata
Large Tortoiseshell
Nymphalis polychloros
Large White
Pieris brassicae
Long-tailed Blue
Lampides boeticus
Lulworth Skipper
Thymelicus acteon
Marbled White
Melanargia galathea
Marsh Fritillary
Euroydryas aurinia
Meadow Brown
Maniola jurtina
Monarch
Danaus plexippus
Mountain Ringlet
Erebia epiphron
Northern Brown Argus
Aricia artaxerxes
Orange Tip
Anthocharis cardamines
Painted Lady
Cynthia cardui
Pale Clouded Yellow
Colias hyale
Peacock
Inachis io
Pearl-bordered Fritillary
Boloria euphrosyne
Purple Emperor
Apatura iris
Purple Hairstreak
Quercusia quercus
Red Admiral
Vanessa atalanta
Ringlet
Aphantopus hyperantus
Scotch Argus
Erebia aethiops
Silver-spotted Skipper
Hesperia comma
Silver-studded Blue
Plebejus argus
Silver-washed Fritillary
Argynnis paphia
Small Blue
Cupido minimus
Small Copper
Lycaena phlaeas
Small Heath
Coenonympha pamphilus
Small Pearl-bordered Fritillary
Boloria selene
Small Skipper
Thymelicus sylvestris
Small Tortoiseshell
Aglais urticae
Small White
Pieris rapae
Speckled Wood
Pararge aegeria
Swallowtail
Papilio machaon
Wall Brown
Lasiommata megera
White Admiral
Ladoga camilla
White-letter Hairstreak
Satyrium (Strymonidia) w-album
Wood White
Leptidea sinapis

Garden plants

Alyssum
Alyssum, genus of 150 spp.
Aster
Callistephus chinensis
Aubretia
Aubretia deltoidea
Bladder Campion
Silene vulgaris
Buddleia
Buddleia davidii and others
Candytuft
Iberis, genus of 30 spp.
Catmint
Nepeta, genus of 250 spp.
Coreopsis
Coreopsis, genus of 120 spp.
Evening Primrose
Oenothera spp.
Feverfew
Chrysanthemum parthenium
Goldenrod
Solidago, genus of 100 spp.
Hebe
Hebe, genus of 100 spp
Honesty
Lunaria, genus of 3 spp.
Honeysuckle
Lonicera periclymenum
Hyssop
Hyssopus officinalis
Ice Plant
Sedum spectabile
Lavender
Lavandula spica
Marigold
Calendula officinalis, Tagetes, genus of 50 spp.
Michaelmas Daisy
Aster, genus of 500 spp.
Nasturtium
Tropaeolum majus
Night Scented Stocks
Matthiola bicornis
Petunia
Petunia, genus of 40 spp.
Phlox
Phlox, genus of 66 spp.
Privet
Ligustrum, genus of 44 spp., eg. L. ovalifolium
Red Valerian
Centranthus ruber
Scabious
Scabiosa, genus of 100 spp.
Sweet Rocket
Hesperis matronalis
Sweet William
Dianthus barbatus
Tobacco Plant
Nicotiana spp.
Verbena
Verbena, genus of 100 spp.
Wallflower
Cheiranthus, genus of 10 spp.
White Jasmine
Jasminum officinale
Zinnia
Zinnia, genus of 20 spp.

Wild flowers & plants

Agrimony
Agrimonia eupatoria
Betony
Betonica officinalis

Bird's Foot Trefoil
Lotus corniculatus
Bitter Vetch
Lathyrus montanus
Black Knapweed
Centaurea nigra
Bladder Senna
Colutea arborescens
Bluebell
Endymion non-scryptus
Borage
Borago officinalis
Bramble
Rubus spp.
Buck's Horn Plantain
Plantago coronopus
Bugle
Ajuga reptans
Cherry
Prunus spp.
Clover
Trifolium spp.
Coltsfoot
Tussilago farfara
Common Cow Wheat
Melampyrum pratense
Common Dog Violet
Viola riviniana
Common Rock Rose
Helianthemum chamaecistus
Common Sorrel
Rumex acetosa
Common Stinging Nettle
Urtica dioica
Common Stork's Bill
Erodium cicutarium
Common Valerian
Valeriana officinalis
Cowslip
Primula veris
Creeping Thistle
Cirsium arvense
Cuckoo Flower
Cardamine pratensis
Dandelion
Taraxacum officinale
Devil's Bit Scabious
Succisa pratensis
Dock
Rumex spp.
Dove's Foot Cranesbill
Geranium molle
Everlasting Pea
Lathyrus latifolius
Fennel
Foeniculum vulgare
Fleabane
Pulicaria vulgaris
Garlic Mustard
Alliaria petiolata
Germander Speedwell
Veronica chamaedrys
Great Bird's Foot Trefoil
Lotus uliginosus
Greater Knapweed
Centaurea scabiosa
Hairy Violet
Viola hirta
Heath
Erica spp.
Heather
Calluna vulgaris
Hemp Agrimony
Eupatorium cannabinum
Hedge Mustard
Sisymbrium officinale
Honesty
Lunaria annua
Honeysuckle
Lonicera periclymenum
Horseshoe Vetch
Hippocrepis comosa
Kidney Vetch
Anthylis vulneraria
Knapweeds
Centaurea spp.
Lady's Smock
Cardamine pratensis
Lucerne
Medicago sativa
Marjoram
Origanum vulgare
Marsh Thistle
Cirsium palustre
Marsh Violet
Viola palustris
Meadowsweet
Filipendula ulmaria
Meadow Vetchling
Lathyrus pratensis
Milk Parsley
Peucedanum palustre
Milkweeds
Asclepias spp.
Ox-Eye Daisy
Chrysanthemum leucanthemum
Primrose
Primula vulgaris
Ragged Robin
Lychnis flos-cuculi
Ragwort
Senecio jacobaea
Red Campion
Silene dioica
Red Clover
Trifolium pratense
Red Valerian
Centranthus ruber
Ribwort Plantain
Plantago lanceolata
Salad Burnet
Poterium sanguisorba
Sheep's Sorrel
Rumex acetosella
Small Nettle
Urtica urens
Snowberry
Symphoricarpos albus
Sowthistle
Sonchus spp.
Spear Thistle
Cirsium vulgare
Teasel
Dipsacus fullonum
Thistles
Cirsium spp. and Carduus spp.
Tormentil
Potentilla erecta
Valerian
Valeriana officinalis
Violets
Viola spp.
Viper's Bugloss
Echium vulgare
Water Dock
Rumex hydrolapanthum
Watermint
Mentha aquatica
Wild Angelica
Angelica sylvestris
Wild Mignonette
Reseda lutea
Wild Strawberry
Fragaria vesca
Wild Thyme
Thymus praecox

Trees and shrubs

Alder Buckthorn
Frangula alnus
Blackthorn
Prunus spinosa
Broom
Sarothamnus scoparius
Buckthorn
Rhamnus catharticus
Buddleia
Buddleia spp.
Crack Willow
Salix fragilis
Currant
Ribes spp.
Dogwood
Thelycrania sanguinea
Elm
Ulmus spp.
Goat Willow
Salix caprea
Gorse
Ulex spp.
Grey Willow
Salix cinerea
Hawthorn
Crataegus spp.
Holly
Ilex aquifolium
Hop
Humulus lupulus
Ivy
Hedera helix
Oak
Quercus spp.
Privet
Ligustrum vulgare
Sallow
Salix spp.
Wych Elm
Ulmus glabra

Wild grasses

Blue Moor Grass
Seslaria caerulea
Bristle Bent
Agrostis setacea
Cock's Foot
Dactylis glomerata
Creeping Soft Grass
Holcus mollis
Early Hair Grass
Aira praecox
False Brome
Brachypodium sylvaticum
Hare's Tail Cotton Grass
Eriophorum vaginatum
Mat Grass
Nardus stricta
Purple Moor Grass
Molinia caerulea
Red Fescue
Festuca rubra
Sheep's Fescue
Festuca ovina
Timothy
Phleum pratense
Tor Grass
Brachypodium pinnatum
Tufted Hair Grass
Deschampsia cespitosa
Upright Brome
Bromus erectus
White-beaked Sedge
Carex curta
Yorkshire Fog
Holcus lanatus

SCIENTIFIC TO ENGLISH NAMES

Aglais urticae
Small Tortoiseshell
Anthocharis cardamines
Orange-tip
Apatura iris
Purple Emperor
Aphantopus hyperantus
Ringlet
Aporia crataegi
Black-veined White
Argynnis adippe
High Brown Fritillary
Argynnis aglaja
Dark Green Fritillary
Argynnis paphia
Silver-washed Fritillary
Aricia agestis
Brown Argus
Aricia artaxerxes
Northern Brown Argus
Boloria euphrosyne
Pearl-bordered Fritillary
Boloria selene
Small Pearl-bordered Fritillary
Callophrys rubi
Green Hairstreak
Carterocephalus palaemon
Chequered Skipper
Celastrina argiolus
Holly Blue
Coenonympha pamphilus
Small Heath
Coenonympha tullia
Large Heath
Colias croceus
Clouded Yellow
Colias hyale
Pale Clouded Yellow
Cupido minimus
Small Blue
Cynthia cardui
Painted Lady
Danaus plexippus
Monarch
Erebia aethiops
Scotch Argus
Erebia epiphron
Mountain Ringlet
Erynnis tages
Dingy Skipper
Euroydryas aurinia
Marsh Fritillary
Gonepteryx rhamni
Brimstone
Hamearis lucina
Duke of Burgundy
Hesperia comma
Silver-spotted Skipper
Hipparchia semele
Grayling
Inachis io
Peacock
Ladoga camilla
White Admiral
Lampides boeticus
Long-tailed Blue
Lasiommata megera
Wall Brown
Leptidea sinapis
Wood White
Lycaena dispar
Large Copper
Lycaena phlaeas
Small Copper
Lysandra bellargus
Adonis Blue
Lysandra coridon
Chalkhill Blue
Maculinea arion
Large Blue
Maniola jurtina
Meadow Brown
Melanargia galathea
Marbled White
Melitaea cinxia
Glanville Fritillary
Mellicta athalia
Heath Fritillary
Nymphalis antiopa
Camberwell Beauty
Nymphalis polychloros
Large Tortoiseshell
Ochlodes venata
Large Skipper
Papilio machaon
Swallowtail
Pararge aegeria
Speckled Wood
Pieris brassicae
Large White
Pieris napi
Green-veined White
Pieris rapae
Small White
Plebejus argus
Silver-studded Blue
Polygonia c-album
Comma
Polyommatus icarus
Common Blue
Pontia daplidice
Bath White
Pyrgus malvae
Grizzled Skipper
Pyronia tithonus
Gatekeeper
Quercusia quercus
Purple Hairstreak
Satyrium (Strymonidia) pruni
Black Hairstreak
Satyrium (Strymonidia) w-album
White-letter Hairstreak
Thecla betulae
Brown Hairstreak
Thymelicus acteon
Lulworth Skipper
Thymelicus lineola
Essex Skipper
Thymelicus sylvestris
Small Skipper
Vanessa atalanta
Red Admiral

INDEX

This index lists English names only. To find the Scientific names of butterflies and plants see pages 75 - 76. For Scientific to English names see page 77.